Electronic-Photonic Integrated Systems for Ultrafast Signal Processing

Johann Christoph Scheytt · Christian Kress ·
Manfred Berroth · Stephan Pachnicke ·
Jeremy Witzens
Editors

Electronic-Photonic Integrated Systems for Ultrafast Signal Processing

Editors
Johann Christoph Scheytt
Schaltungstechnik
Heinz Nixdorf Institut
Paderborn, Germany

Christian Kress
Schaltungstechnik
Heinz Nixdorf Institut
Paderborn, Germany

Manfred Berroth
Institut für Elektrische und Optische
Nachrichtentechnik (INT)
Universität Stuttgart
Stuttgart, Baden-Württemberg, Germany

Stephan Pachnicke
Nachrichtenübertragungstechnik
Christian-Albrechts-Universität zu Kiel
Kiel, Germany

Jeremy Witzens
Institut und Lehrstuhl für Integrierte Photonik
RWTH Aachen University
Aachen, Nordrhein-Westfalen, Germany

ISBN 978-3-032-08339-5   ISBN 978-3-032-08340-1 (eBook)
https://doi.org/10.1007/978-3-032-08340-1

This work was supported by Universität Paderborn.

© The Editor(s) (if applicable) and The Author(s) 2026. This book is an open access publication.

Open Access This book is licensed under the terms of the Creative Commons Attribution 4.0 International License (http://creativecommons.org/licenses/by/4.0/), which permits use, sharing, adaptation, distribution and reproduction in any medium or format, as long as you give appropriate credit to the original author(s) and the source, provide a link to the Creative Commons license and indicate if changes were made.
The images or other third party material in this book are included in the book's Creative Commons license, unless indicated otherwise in a credit line to the material. If material is not included in the book's Creative Commons license and your intended use is not permitted by statutory regulation or exceeds the permitted use, you will need to obtain permission directly from the copyright holder.
The use of general descriptive names, registered names, trademarks, service marks, etc. in this publication does not imply, even in the absence of a specific statement, that such names are exempt from the relevant protective laws and regulations and therefore free for general use.
The publisher, the authors and the editors are safe to assume that the advice and information in this book are believed to be true and accurate at the date of publication. Neither the publisher nor the authors or the editors give a warranty, expressed or implied, with respect to the material contained herein or for any errors or omissions that may have been made. The publisher remains neutral with regard to jurisdictional claims in published maps and institutional affiliations.

This Springer imprint is published by the registered company Springer Nature Switzerland AG
The registered company address is: Gewerbestrasse 11, 6330 Cham, Switzerland

If disposing of this product, please recycle the paper.

Preface

Optical signal processing offers significant advantages in terms of speed (data rate and bandwidth) and is generally much more energy-efficient than electrical signal processing. Furthermore, optical signal transmission involves less loss, which is why high-speed fiber-optic communication networks form the "backbone" of the internet. In addition to these benefits, optical oscillators (lasers and frequency combs) have fundamentally superior spectral properties compared to their electronic counterparts. Conversely, electronics are highly cost-efficient, enable sophisticated signal processing and use extremely small processing elements (transistors), as well as being programmable by means of software.

Currently, optical and electronic circuits tend to be separate components within a system, with optical circuits containing only a few components. However, in recent years, photonic integration technologies such as silicon photonics, indium phosphide, silicon nitride and thin-film lithium niobate technology have advanced significantly. Photonic integration technology enables large numbers of nano-scale optical devices to be combined with complex nano-scale electronics. Consequently, complex optical signal processing can be combined with analogue electronics, digital processors, memory and software. Miniaturized electronic-photonic systems can be integrated onto a single chip or onto a few chips within a single package. Photonic integration therefore allows for miniaturized optics and the close proximity of optics and electronics, reducing energy consumption and size, as well as enabling programmable optics and higher optical system complexity at low cost. These new possibilities break up the paradigm of separate optical and electronic signal processing domains and make it necessary to fundamentally rethink how signal processing algorithms, signal processors, communication networks and sensors should be optimally realized in order to fully exploit the potential of integrated nanophotonic/nanoelectronic systems.

About the SPP 2111

This book describes recent progress in selected fields of this exciting and fast evolving field. The presented research results are outcomes of a research framework program at Deutsche Forschungsgemeinschaft (DFG), called "Electronic-Photonic Integrated Systems for Ultrafast Signal Processing" (SPP 2111). The program started in 2018 and ended in 2025. In the program researchers from electrical engineering, physics and computer science conducted interdisciplinary research into novel electronic-photonic integrated systems.

The programme concept was developed by Prof. Manfred Berroth (University of Stuttgart), Prof. Norbert Hanik (Technical University of Munich), Prof. Franz-Xaver Kärtner (University of Hamburg and Deutsches Elektronen-Synchrotron (DESY)), Prof. J. Christoph Scheytt (Paderborn University and Coordinator) and Prof. Lars Zimmermann (Technical University of Berlin), and was submitted to the Deutsche Forschungsgemeinschaft (DFG) in late 2016. The programme began with a nationwide call for proposals in 2017 for an initial three-year term. A second call for proposals was published in 2021 for a second three-year term. 21 projects were funded, providing a total of 41 Ph.D. student positions. Members of SPP 2111 produced around 215 publications, which were cited 2330 times at this point.

About This Book

The chapters in this volume summarize the results of projects carried out within the program.

Chapter 1, presents research into broadband wavelength conversion using optimized multimodal optical waveguides with high cubic nonlinearity. The first part explains the physical principles and mathematical modelling of ultra-broadband optical wavelength conversion in a highly nonlinear multimodal silicon-on-insulator (SOI) waveguide. Examples of optimized SOI waveguide designs that enable ultra-broadband conversion of signals within or between various wavelength bands of an optical fibre are presented. The general process of fabricating SOI waveguides is outlined, and the measured key parameters of samples from various generations of optimized waveguides are presented. The performance of the developed optical converters is demonstrated in laboratory experiments.

Chapter 2, describes the development and experimental validation of an optic-electronic-optic (OEO) interferometer that supports flexible coherent adddrop operations with a high bandwidth. The system uses coherent detection, realtime digital signal processing and in-phase/quadrature electro-optic remodulation in a single interferometer arm to allow programmable manipulation of the optical field. Successful demonstration of add-drop operation for a quadrature phase-shift keying (QPSK) signal is presented. In addition to single-channel demonstrations, the authors present experimental results using intradyne detection in a densewavelengthdivision multiplexing (WDM) scenario, demonstrating scalability to multi-carrier systems. The development and experimental validation of various indium phosphidebased integrated OEO interferometer photonic integrated circuits and modules are discussed.

Chapter 3, presents recent research into ultra-broadband photonically assisted analog-to-digital converters (ADCs). This work covers both time- and frequency-domain approaches for photonically assisted ADCs, and explores how these architectures could be expanded to incorporate additional signal processing tasks, such as equalization. Optically triggered track-and-hold amplifiers with an equivalent jitter below 80 fs RMS in a signal frequency range of 20–70 GHz are shown. A frequency-domain architecture implementing optical arbitrary waveform measurement is explained. Design, implementation and experimental validation of the architecture up to record signal bandwidths of 610 GHz is discussed.

Chapter 4, describes an electro-optical digital-to-analogue converter (DAC). This concept uses Nyquist pulse sequences in the optical domain to generate an optical pulse sequence with very high precision. Since the pulses can be intensity-modulated using time multiplexing, high sampling rates can be generated with comparatively low electronic bandwidth. The chapter describes the mathematical theory, circuit architectures, and measured results of critical components of the Nyquist Pulse Synthesizer DAC.

In Chap. 5, progress towards the design and fabrication of an integrated, low-jitter MLL and its applications is presented. First, the origin and scaling of timing jitter in ultra-short-pulse MLLs is reviewed and compared to microwave oscillator timing jitter. Then, architectural strategies for jitter suppression in chip-scale MLLs are discussed and recent experimental progress towards integrated MLLs is presented. A major application of integrated low-jitter MLLs will be in analogue-to-digital converters (ADCs), as the precision of high-speed ADCs is currently severely limited by sampling clock jitter. Consequently, an ultrabroadband photonic ADC application is presented that uses an MLL signal as the sampling clock. Finally, for applications requiring a free spectral range in the frequency domain wider than can be achieved at chip scale, the chapter presents an implementation of an on-chip filter using CMOS-compatible silicon nitride technology that can be cascaded with the MLL on the same photonic integrated circuit.

Finally, in the last chapter (Chap. 6), the concept of a nonlinear frequency division multiplexed transmission system is presented, in which a broadband nonlinear spectrum is stitched together by means of a photonic integrated circuit. Four independently modulated wavelength-division-multiplexed channels are linearly multiplexed together with trapezoidal filtering and partially overlapping spectra, approximating a seamless wideband nonlinear spectrum in which guard bands are avoided. It is demonstrated that this approach mitigates the problem of nonlinear interaction between the channels, enabling more efficient use of the spectrum. Various forms of nonlinear spectrum modulation are discussed, and the system is simulated to determine its performance limits.

As coordinator of the program, I would like to thank all the authors who submitted their work and the co-editors for their helpful comments. I would also like to acknowledge the DFG for accepting and sponsoring the priority program SPP 2111 on Electronic-Photonic Integrated Systems for Ultrafast Signal Processing. Dr. Damian Dudek and Dr. Bastian Mohr from DFG provided important support to the program. I would like to convey special thanks to them.

It was a great pleasure to coordinate the programme. It was also wonderful to witness excellent yet abstract ideas being transformed into tangible results.

Paderborn, Germany Johann Christoph Scheytt
June 2025

Contents

1 **Ultra-Broadband Photonic Signal-Processor** 1
Norbert Hanik, Tasnad Kernetzky, Ulrike Höfler, Yizhao Jia,
Colja Schubert, Masoumeh Karvar, Isaac Sackey, Ronald Freund,
Gregor Ronniger, and Lars Zimmermann

2 **Optic-Electronic-Optic Interferometers for Ultrabroadband Arbitrary Digital Signal Processing** 51
Sebastian Randel, Md. Salek Mahmud, Alexander Schindler,
Patrick Runge, and Martin Schell

3 **Ultra-Broadband Photonically Assisted Analog-to-Digital-Converters** 83
Jeremy Witzens, Daniel Drayss, Dengyang Fang, Alvaro Moscoso Mártir,
Juliana Müller, Maxim Weizel, Andrea Zazzi, Wolfgang Freude,
Christian Koos, Sebastian Randel, and J. Christoph Scheytt

4 **Precise Optical Nyquist Pulse Synthesizer Digital-to-Analog Converter** 119
J. Christoph Scheytt, Tobias Schwabe, Karanveer Singh, Christian Kress,
and Thomas Schneider

5 **Integrated Low Jitter Mode-Locked Lasers and Pulse/Spectrum Shapers** 145
Jeremy Witzens, Milan Sinobad, Tengizi Abramishvili, Pascal Gehrmann,
Andrea Zazzi, Mike Külkens, Jan Lorenzen, Alvaro Moscoso Mártir,
Neetesh Singh, and Franz X. Kärtner

6 **Stitched-Spectrum Nonlinear Frequency Division Multiplexed Transmission Systems Using Photonic Integration** 181
Stephan Pachnicke, Olaf Schulz, Alvaro Moscoso-Mártir,
and Jeremy Witzens

Ultra-Broadband Photonic Signal-Processor

Norbert Hanik, Tasnad Kernetzky, Ulrike Höfler, Yizhao Jia, Colja Schubert, Masoumeh Karvar, Isaac Sackey, Ronald Freund, Gregor Ronniger and Lars Zimmermann

Abstract

Flexible, ultra-broadband, optical wavelength conversion will be one of the key issues in future optical multiband networks that make use of the full low-attenuation bandwidth of Standard Single-Mode Fibers (SSMF). This chapter outlines the authors' work to realize broadband wavelength conversion between extreme optical wavelength bands using optimized multimodal optical waveguides with high cubic nonlinearity. In a first part, the physical principles of ultra-broadband optical wavelength conversion in a highly nonlinear multimodal Silicon-On-Insulator (SOI) waveguide, and methods to model and optimize its functionality, are explained. Subsequently, examples of optimized SOI waveguide designs are presented that enable ultra-broadband conversion of signals within, or between various low-attenuation wavelength bands of an optical fiber. In a second paragraph, the fabrication of SOI waveguides is delineated in general, and measured key parameters of samples of various generations of optimized waveguides are revealed.

N. Hanik (✉) · T. Kernetzky · U. Höfler · Y. Jia
TU München, Munich, Germany
e-mail: norbert.hanik@tum.de

C. Schubert · M. Karvar · R. Freund
Fraunhofer Gesellschaft Heinrich Hertz-Institut (HHI), Berlin, Germany

L. Zimmermann
Institut für Halbleiterphysik und Photonik (IHP), Frankfurt (Oder), Germany

I. Sackey
Infinera, Ottawa, ON, Canada

G. Ronniger
Institut für Mikroelektronik, Stuttgart, Germany

© The Author(s) 2026
J. C. Scheytt et al. (eds.), *Electronic-Photonic Integrated Systems for Ultrafast Signal Processing*, https://doi.org/10.1007/978-3-032-08340-1_1

In a final section, the performance of the developed optical converters is demonstrated exemplarily in laboratory experiments.

Keywords

Optical communications · Multi-mode waveguides · Wavelength conversion · Four-wave mixing

Acronyms

ASE	Amplified Spontaneous Emission Noise
AWGN	Additive White Gaussian Noise
AOWC	All-Optical Wavelength Converter
B2B	Back-to-Back
BER	Bit Error Rate
BS	Bragg Scattering
CD	Chromatic Dispersion
CE	Conversion Efficiency
CMOS	Complementary Metal Oxide Semiconductor
CW	Continuous Wave
DAC	Digital-to-Analog Converter
DBP	Digital Back Propagation
DEMUX	Demultiplexer
DSP	Digital Signal Processing
DWDM	Dense Wavelength Division Multiplexing
ECL	External Cavity Laser
EDFA	Erbium-Doped Fiber Amplifier
EPIC	Electronic-Photonic Integrated Circuit
FCA	Free Carrier Absorption
FDM	Finite Difference Method
FWM	Four-Wave Mixing
GC	Grating Coupler
HNLF	Highly Nonlinear Fiber
IL	Insertion Loss
IQ	In-Phase and Quadrature
KK-RX	Kramers-Kronig Receiver
MADM	Mode Add-Drop Multiplexer
MCF	Multi-Core Fiber
MMF	Multi-Mode Fiber
MMWG	Multi-Mode Waveguide

MUX	Multiplexer
NR	Nano-Rib
OPC	Optical Phase Conjugation
OPC-FWM	Optical Phase Conjugation Based on Four-Wave Mixing
OSA	Optical Spectrum Analyzer
OSNR	Optical Signal-to-Noise Ratio
QPSK	Quadrature Phase-Shift Keying
PC	Polarization Controller
PD	Photodiode
PDFA	Praseodymium-Doped Fiber Amplifier
PDLA	Photonic Dispersion and Loss Analyzer
PDM	Polarization-Division Multiplexing
PIC	Photonic Integrated Circuit
PM	Phase Matching
QAM	Quadrature Amplitude Modulation
SE	Spectral Efficiency
SMF	Single-Mode Fiber
SNR	Signal-to-Noise Ratio
SOI	Silicon-On-Insulator
SPM	Self-Phase Modulation
SRS	Stimulated Raman Scattering
SSMF	Standard Single-Mode Fiber
SDM	Space-Division Multiplexing
SP	Single-Polarization
TADC	Tapered Asymmetric Directional Coupler
TE	Transverse Electric
TPA	Two-Photon Absorption
T-OBPF	Tunable Optical Band-Pass Filter
VOA	Variable Optical Attenuator
WDM	Wavelength-Division Multiplexing
WG	Waveguide
WLC	Wavelength Conversion
WSS	Wavelength-Selective Switch
XT	Cross-Talk

1.1 Introduction

With the availability of mature, low-attenuation optical Single-Mode Fiber (SMF) in the 1980s all segments of the world-wide communications network have been transformed from copper to optics, leading to a tremendous boost in transmission capacity. Since then, optical communications networks are continuously upgraded to keep pace with the dynamical growth of the world-wide data load. This traffic increase, typically between 25 and 80% per year [2], is not expected to saturate in the mid-future. Network operators keep up with this evolution by exploiting so-far unused degrees of freedom, e.g. by extensively modulating in-phase and quadrature components of optical carriers up to 1024-ary Quadrature Amplitude Modulation (QAM) [3], and by applying capacity-achieving coded modulation schemes with probabilistic shaping [4, 5]. Currently, the optical C-band, which exhibits minimum signal attenuation, is fully equipped with Wavelength-Division Multiplexing (WDM) optical channels, approaching ultimate (Shannon-) spectral efficiency. Systems that employ the adjacent L-band are being installed. Exploiting the lower-wavelength regions of the O-, E-, and S-band (Fig. 1.1), is the straightforward option to gain additional transmission-capacity. In the long term, Space-Division Multiplexing (SDM) using either Multi-mode Fibers (MMFs) or Multi-Core Fibers (MCFs), or a combination of both, are identified as potential solutions to satisfy a further capacity increase in fiber-optic communication networks [2]. Exploiting the full bandwidth of Standard Single-Mode Fiber (SSMF) according to Fig. 1.1 raises the demand for efficient devices that achieve broadband switching of multiples of wavelength-channels between wavelength bands to enable full flexibility of these multi-band optical networks. Conventional electronics, which is unable to process data overa multi-terahertz bandwidth, is not an option in the foreseeable future. However,

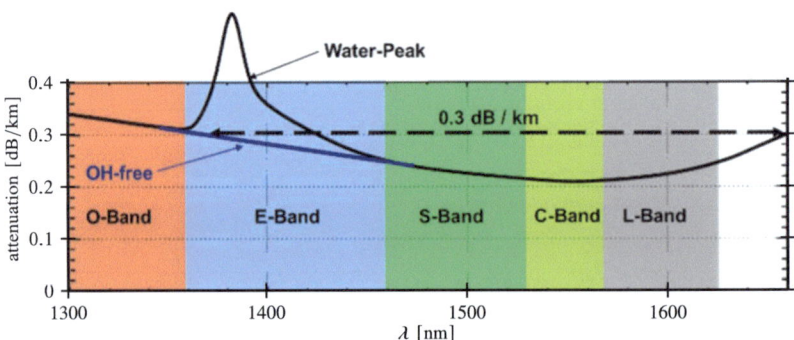

Fig. 1.1 Low-attenuation infrared wavelength bands of a Single-Mode Fiber [1]

a promising solution is all-optical wavelength conversion, utilizing the cubic nonlinearity of several optical materials. While single-mode devices exhibit an obstructive bandwidth limit due to the high phase mismatch of the underlying Four-Wave Mixing (FWM) process, it seemed to be possible to achieve higher bandwidths via nonlinear interaction of more than one propagating mode [6]. In this chapter, a survey of the authors' work in the framework of the DFG-Priority-Programme SPP2111: *"Electronic-Photonic Integrated Systems for Ultra-Fast Signal Processing"* to design, optimize, realize, and evaluate multimodal SOI waveguide structures that enable optical ultra-broadband wavelength conversion between arbitrary low-attenuation wavelength bands of silica fibers is given. First, we introduce the mathematical theory of guided optical waveguide modes and the appropriate modeling of linear and nonlinear perturbations that affect their propagation. The underlying multimodal FWM process and applied methods to optimize the waveguide design are outlined. Three samples of optimized waveguides are numerically evaluated, showing unique properties to convert signal wavelengths between dedicated optical λ-bands. Several generations of optimized waveguides have then been fabricated and characterized; results are given in a second paragraph. Finally, the performance of selected samples has been evaluated in laboratory experiments. Exemplarily, successful ultra-broadband C-to-O-band conversion and broadband Optical Phase Conjugation (OPC) are presented in this concluding section.

1.2 Mathematical Modeling of Optical Waveguide Modes and Propagation

1.2.1 Guided Modes in SOI-Waveguides

The optical waveguide under investigation is composed of a silicon core and a silica cladding that surrounds that core. The refractive index n_{core} of the core is increased compared to the refractive index n_{clad} of the cladding. The effect of this **refractive index profile** is, that incident light of appropriate frequency is guided within the waveguide. The transition of the refractive index from the core to the cladding region can be rectangular –so-called step-index profile– or be governed by a more complex function. By appropriate design some transmission properties of the waveguide can be controlled. The distribution of light intensity across the core (transversal field distribution) is dependent on the core diameter, on the refractive-index profile, and on the wavelength of the propagating light. Due to their fundamentally different properties Single-Mode and Multi-Mode Waveguides (MMWGs) are distinguished. First of all, **Maxwell's equations** govern the interaction of electrical and magnetic fields and serve as a starting point:

$$\operatorname{curl} \vec{E} = -\frac{\partial \vec{B}}{\partial t} \tag{1.1}$$

$$\operatorname{curl} \vec{H} = \frac{\partial \vec{D}}{\partial t} \tag{1.2}$$

$$\operatorname{div} \vec{D} = 0 \tag{1.3}$$

$$\operatorname{div} \vec{B} = 0 \tag{1.4}$$

In the above equations, $\vec{E}$ and $\vec{H}$ denote the electrical and magnetic field strengths, having x-, y-, and z-components in our Cartesian coordinate system. $\vec{D}$ and $\vec{B}$ are the electric and magnetic flux densities. The **material equations** consider the specific properties of the wave-guiding material:

$$\vec{D} = \epsilon_0 \cdot \boldsymbol{\epsilon}_r \cdot \vec{E} + \vec{P}^{\mathrm{nl}} \tag{1.5}$$

$$\vec{B} = \mu_0 \cdot \vec{H}, \tag{1.6}$$

where ϵ_0 is the vacuum permittivity, μ_0 the vacuum permeability and $\vec{P}^{\mathrm{nl}}$ the nonlinear material polarization. The matrix $\boldsymbol{\epsilon}_r$ is the so-called permittivity of the medium which is now decomposed into three contributions:

$$\boldsymbol{\epsilon}_r = \boldsymbol{I} + \frac{\boldsymbol{\chi}^{[1]}}{\epsilon_0} - \mathrm{j}\epsilon_r'' \boldsymbol{I} = \epsilon_r' \boldsymbol{I} + \boldsymbol{\delta\epsilon}_r - \mathrm{j}\epsilon_r'' \boldsymbol{I} \tag{1.7}$$

with the identity matrix $\boldsymbol{I}$, the linear susceptibility matrix $\boldsymbol{\chi}^{[1]}$ and

$$\boldsymbol{\delta\epsilon}_r = \begin{pmatrix} \delta\epsilon_{r,xx} & \delta\epsilon_{r,xy} & \delta\epsilon_{r,xz} \\ \delta\epsilon_{r,yx} & \delta\epsilon_{r,yy} & \delta\epsilon_{r,yz} \\ \delta\epsilon_{r,zx} & \delta\epsilon_{r,zy} & \delta\epsilon_{r,zz} \end{pmatrix}. \tag{1.8}$$

Here, $\epsilon_r' = n^2(\omega, x, y)$ models the reduced light velocity in the medium and pulse spreading due to material dispersion, where n is the position-dependent refractive index of the material and x and y are the transversal coordinates of the waveguide as the guided lightwave propagates along the z-direction. The matrix $\boldsymbol{\delta\epsilon}_r$ includes all linear perturbations of the waveguide, leading to linear coupling of the guided modes. The imaginary part ϵ_r'' represents the linear attenuation of the waveguide. Rearranging all perturbations into one vector $\vec{P}'$ with

$$\vec{P}' = -\mathrm{j}\epsilon_0 \cdot \epsilon_r'' \vec{E} + \epsilon_0 \cdot \boldsymbol{\delta\epsilon}_r \vec{E} + \vec{P}^{\mathrm{nl}} \tag{1.9}$$

yields a modified equation for the electric flux density according to

$$\vec{D} = \epsilon_0 \cdot \epsilon_r' \vec{E} + \vec{P}'. \tag{1.10}$$

The electric flux density $\vec{D}$ consists now of a contribution reflecting the *ideal* waveguide properties modeled by ϵ_r' and a second contribution $\vec{P}'$ including all perturbations ϵ_r'', $\boldsymbol{\delta\epsilon}_r$

and the cubic perturbation due to $\vec{P}^{nl}$. With some basic mathematical transformations, a first version of the wave equation is derived

$$\Delta \vec{E} - \mathrm{grad}\left(\mathrm{div}\,\vec{E}\right) = \mu_0 \cdot \frac{\partial^2}{\partial t^2}\vec{D}\,. \tag{1.11}$$

Evaluating the electric flux density field gives the equation

$$\mathrm{div}\,\vec{D} = \epsilon_0 \cdot \vec{E} \cdot \mathrm{grad}\,\epsilon_r' + \epsilon_0 \cdot \epsilon_r' \cdot \mathrm{div}\,\vec{E} + \mathrm{div}\,\vec{P}' \\ = 0. \tag{1.12}$$

Applying few more mathematical operations the general wave equation can be stated as

$$\Delta \vec{E} + \underbrace{\mathrm{grad}\left(\frac{1}{\epsilon_r'} \cdot \vec{E} \cdot \mathrm{grad}\,\epsilon_r' + \frac{1}{\epsilon_0 \epsilon_r'} \cdot \mathrm{div}\,\vec{P}'\right)}_{\text{\textcircled{A}}} - \mu_0 \cdot \epsilon_0 \cdot \epsilon_r' \cdot \frac{\partial^2}{\partial t^2}\vec{E} = \mu_0 \cdot \frac{\partial^2}{\partial t^2}\vec{P}'\,.$$

$$\tag{1.13}$$

In this vectorial equation the material permittivity $\epsilon_r'(x, y) = n^2(x, y)$ models the transversal structure of the waveguide. As the waveguide is much longer than the propagating signal's wavelength, the z-dependence is neglected. Thus, $n(x, y)$ determines the solutions of the ideal wave equation with $\vec{P}'$ being zero. All perturbations of the ideal (lossless) waveguide are included in the material polarization $\vec{P}'$. As we treat the influence of $\vec{P}'$ as very small perturbation of the ideal waveguide properties, we first neglect it and evaluate the ideal wave equation

$$\Delta \vec{E} + \mathrm{grad}\left(\frac{1}{\epsilon_r'} \cdot \vec{E} \cdot \mathrm{grad}\,\epsilon_r'\right) - \mu_0 \cdot \epsilon_0 \cdot \epsilon_r' \cdot \frac{\partial^2}{\partial t^2}\vec{E} = 0. \tag{1.14}$$

This is the differential equation that governs the propagation of a light wave with specific frequency f in an ideal optical waveguide. Ideal means that waveguide effects generating attenuation and linear and nonlinear perturbations are neglected in this equation, by setting $\vec{P}'$ equal to zero. In general, this ideal wave equation must be solved numerically. The resulting guided transversal field distributions $\vec{\Psi}_{\omega_a}^{(m)}(x, y)$ (the eigenfunctions called *modes*), and their corresponding propagation constants $\beta_{\omega_a}^{(m)}$ (the eigenvalues) at frequency ω_a and mode m define the total propagating unmodulated field, given by[1]

$$\vec{E}(x, y, z, t) = \Re\{\underline{\vec{E}}(x, y, z, t)\} \\ = \frac{1}{2}\sum_{a,m} \hat{E}_{\omega_a}^{(m)}(z) \cdot \vec{\Psi}_{\omega_a}^{(m)}(x, y) \cdot e^{j\left(\omega_a t - \beta_{\omega_a}^{(m)} z\right)} + c.c., \tag{1.15}$$

---

[1] The notation indicates the analytic signal, i.e., the complex-valued representation of the real physical field containing only positive frequency components.

where the summation extends over all frequency components a and mode indices m with the complex amplitudes $\hat{E}_{\omega_a}^{(m)}$. The number of guided modes is dependent on the waveguide design and the investigated frequency $\omega = 2\pi f$. This general theory is now applied to our concrete SOI-waveguide [6]. Only for special waveguide geometries the ideal wave equation has analytical solutions. To be able to evaluate arbitrary waveguide profiles, numerical algorithms have to be used [7]. All results presented in the following have been computed with a solver based on a full-vectorial Finite Difference Method (FDM) [8]. The general idea of a FDM is to discretize the waveguide's cross-section into a grid, and approximate differential operators by truncated Taylor series expansions of the fields on the grid points. Finally, a matrix-eigenvalue equation has to be solved numerically, where the eigenvectors are the mode fields and the corresponding eigenvalues represent the associated propagation constants. We make use of the fact that only two of the six (electric and magnetic) field components have to be found, the remaining components are linearly dependent and can be derived from these two. The precision of the calculated solution depends on the grid point density, which itself determines computation time and required memory. The investigated regular silicon Nano-Rib (NR) waveguide is depicted in Fig. 1.2. The core is composed of a slab, rib, and rails. It is surrounded by the silica substrate and the coating. During manufacturing, silicon nitride and silica protective layers are added above the core. The rails on the left and right of the slab are to apply a reverse bias voltage to reduce absorption, generated by the free carriers of the silicon material. Since this voltage does not affect the propagating field, and since the rails are far enough from the rib that they likewise do not affect the modes, they are not included in the simulation. The four lowest order modes of the regular waveguide are shown in Fig. 1.3a–d. All modes are approximately linear polarized and are labeled TE_0, TE_1, TE_2, and TE_3, TE stands for *Transverse Electric*. The index counts the polarity inversions along the x-axis being 0 to 3 for the respective mode.

One additional degree of freedom in the waveguide design is to etch dips into it, which alter the waveguide's dispersion and might be beneficial for the FWM process, as we will

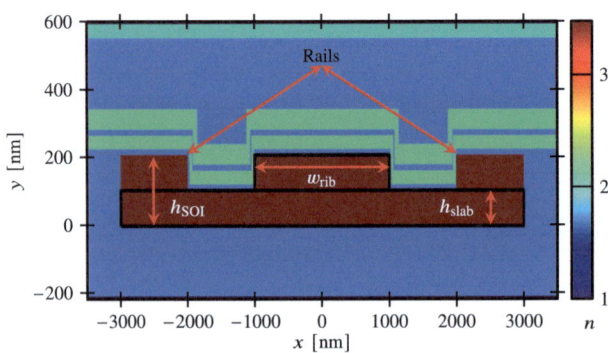

Fig. 1.2 Refractive index profile of the investigated regular Nano-Rib waveguide with layer stack and rails. Note the different scaling of the axis

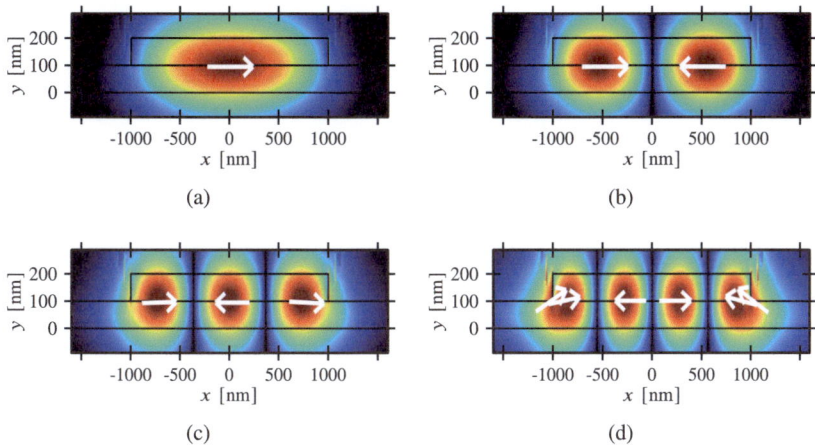

Fig. 1.3 Transversal electrical field distribution of the four lowest-order modes $TE_0 - TE_3$. Colors encode magnitudes (large: red to small: blue). Arrows visualize the direction of light polarization

see later. Parameters are the number of etched dips, and their (identical) width and height, as depicted in Fig. 1.4. Etching dips into the waveguide will mostly affect modes that exhibit significant power at the dip positions. The mode fields in Fig. 1.5a–d clearly show the impact of the two dips on the modes TE_0, TE_2 and TE_3, while they have only little effect on TE_2. Despite its misleading name, the propagation *constants* $\beta^{(m)}$ in Eq. (1.15) are frequency dependent. They are usually expanded into their Taylor series around a center frequency ω_0 according to

$$\beta^{(m)}(\omega) = \beta_0^{(m)} + \left.\frac{\partial \beta^{(m)}(\omega)}{\partial \omega}\right|_{\omega_0} \cdot \Delta\omega \\ + \frac{1}{2} \left.\frac{\partial^2 \beta^{(m)}(\omega)}{\partial \omega^2}\right|_{\omega_0} \cdot \Delta\omega^2 + \frac{1}{6} \left.\frac{\partial^3 \beta^{(m)}(\omega)}{\partial \omega^3}\right|_{\omega_0} \cdot \Delta\omega^3 + \cdots \tag{1.16}$$

Fig. 1.4 Dips etched into the waveguide with parameters h_{dip} and w_{dip}

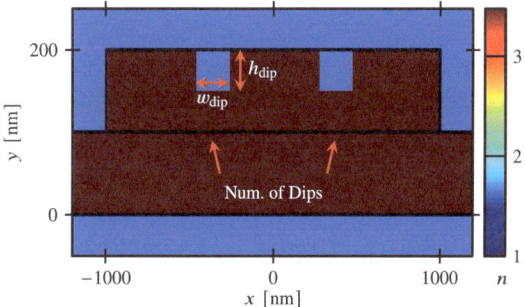

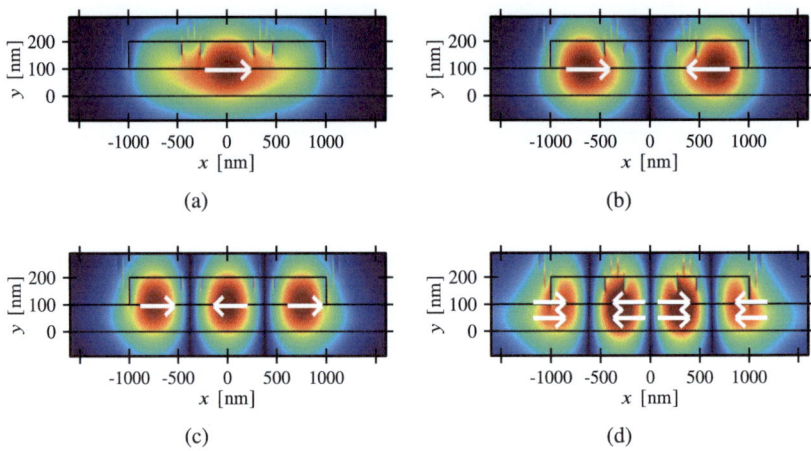

Fig. 1.5 Transversal electrical field distribution of the four lowest-order modes $TE_0 - TE_3$ of a waveguide with two dips

with $\Delta\omega = \omega - \omega_0$. Every addend in Eq. (1.16) has a specific physical impact on the propagating wave: $\beta_0^{(m)}$ is responsible for the phase retardation of an unmodulated optical carrier. The first order derivative

$$\bar{\tau}_{gr}^{(m)}(\omega) = \frac{\partial \beta^{(m)}(\omega)}{\partial \omega} = \frac{\partial \beta^{(m)}(\lambda(\omega))}{\partial \lambda} \frac{\partial \lambda}{\partial \omega} \quad (1.17)$$

models the normalized group delay of the modulated amplitude. The chromatic (group-delay) dispersion is determined by

$$D^{(m)}(\lambda) = \frac{\partial \bar{\tau}_{gr}^{(m)}(\lambda)}{\partial \lambda} = \frac{\partial \bar{\tau}_{gr}^{(m)}(\omega(\lambda))}{\partial \omega} \frac{\partial \omega}{\partial \lambda} = \frac{\partial^2 \beta^{(m)}(\omega)}{\partial \omega^2} \frac{\partial \omega}{\partial \lambda}, \quad (1.18)$$

where $\frac{\partial \omega}{\partial \lambda} = -\frac{2\pi c_0}{\lambda^2}$ and $\frac{\partial \lambda}{\partial \omega} = -\frac{2\pi c_0}{\omega^2}$ with c_0 being the light velocity in vacuum. The third order derivative in Eq. (1.16) models cubic dispersion; higher order derivatives are negligible for typical applications in communications. The Chromatic Dispersion (CD) D itself is composed by the material dispersion, i.e., the frequency dependence of the refractive index of the waveguide material, and the waveguide dispersion, which is caused by the waveguiding effect. Material dispersion can be easily measured for all materials the waveguide is composed of, and is typically given by Sellmeier-equations [6]. Waveguide dispersion, however, highly depends on the waveguide geometry and can be derived from the solutions of the propagating modes.

Figure 1.6 shows group delay and dispersion curves of a regular waveguide and a waveguide with two dips. Comparing the dispersion curves of the TE_3-mode in plot (b) and (d), the effect of the dips is clearly visible, while the TE_2 curve does almost not change. In this

1 Ultra-Broadband Photonic Signal-Processor   11

way, adding dips can be used for fine-tuning the waveguide dispersion and improve Phase Matching (PM) for selected modes.

1.2.2 Four-Wave Mixing

In a Four-Wave Mixing process, four waves propagating in a nonlinear medium are coupled due to the cubic nonlinear susceptibility of the material. Though being detrimental in long-distance communication links, it is beneficial for optical wavelength conversion that we want to realize. The effect of FWM is manifold [9, 10]. However, the concrete processes that we are interested in are Bragg Scattering (BS) and OPC. Our FWM-setup consists of a signal (S) and two pump waves P_1 and P_2 at their respective angular frequencies ω_S, ω_{P_1} and ω_{P_2}, coupled into the nonlinear medium. The frequency of the fourth wave generated by FWM, the so-called idler (I), is determined by the conservation of the total energy of the

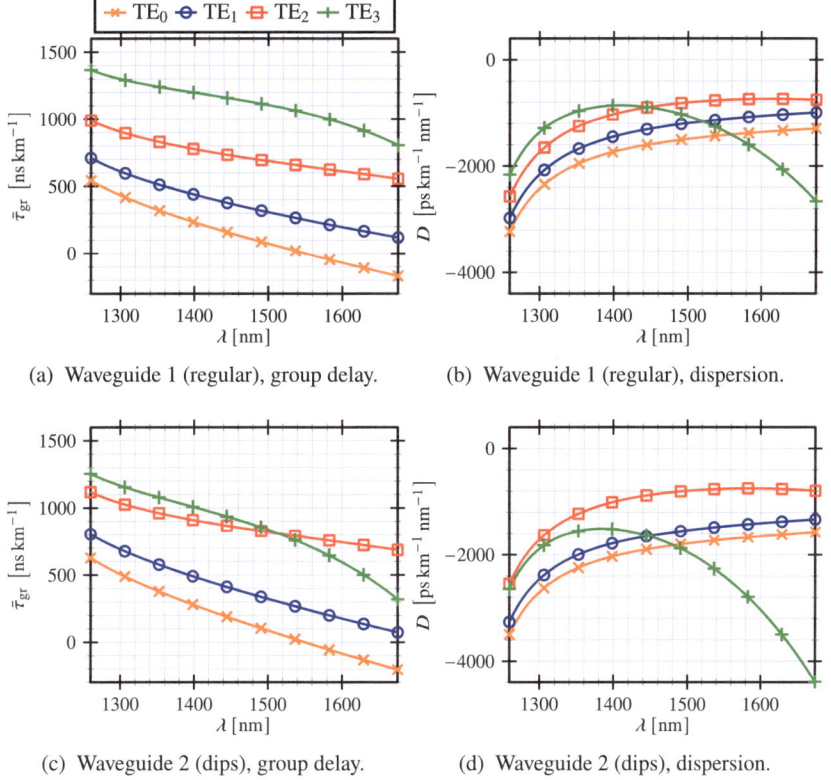

Fig. 1.6 Normalized group delay and dispersion of the four lower-order modes in a regular waveguide (waveguide 1) and a waveguide with two dips (waveguide 2)

interacting photons. This leads to the frequency relation of

$$\hbar\omega_I = \hbar\omega_{P_1} + \hbar\omega_S - \hbar\omega_{P_2} \quad \text{for BS,} \tag{1.19}$$

$$\text{and} \quad \hbar\omega_I = \hbar\omega_{P_1} - \hbar\omega_S + \hbar\omega_{P_2} \quad \text{for OPC.} \tag{1.20}$$

However, not only photon energy, but also the total photon momentum has to be conserved during the FWM-process, which leads to the following relations between the phase constants of the interacting waves:

$$\hbar\beta^{(D)}(\omega_I) = \hbar\beta^{(A)}(\omega_{P_1}) + \hbar\beta^{(B)}(\omega_S) - \hbar\beta^{(C)}(\omega_{P_2})$$
$$\Rightarrow \Delta\beta = \beta^{(A)}(\omega_{P_1}) + \beta^{(B)}(\omega_S) - \beta^{(C)}(\omega_{P_2}) - \beta^{(D)}(\omega_I), \quad \text{for BS} \tag{1.21}$$

and

$$\hbar\beta^{(D)}(\omega_I) = \hbar\beta^{(A)}(\omega_{P_1}) - \hbar\beta^{(B)}(\omega_S) + \hbar\beta^{(C)}(\omega_{P_2})$$
$$\Rightarrow \Delta\beta = \beta^{(A)}(\omega_{P_1}) - \beta^{(B)}(\omega_S) + \beta^{(C)}(\omega_{P_2}) - \beta^{(D)}(\omega_I), \quad \text{for OPC.} \tag{1.22}$$

The letters A, B, C, and D are mode indices, $\Delta\beta$ is the so-called phase mismatch. In principle, for either of the involved frequencies, the laser-to-mode association is a degree of freedom. However, as $\beta_0^{(m)}$ in Eq. (1.16) is subject to fluctuations due to manufacturing imperfections or external perturbations of the waveguide (and should therefore be canceled out to enable PM), only three options for BS and OPC remain:

$$\Delta\beta = \beta^{(A)}(\omega_{P_1}) + \beta^{(A)}(\omega_S) - \beta^{(A)}(\omega_{P_2}) - \beta^{(A)}(\omega_I),$$
$$\Delta\beta = \beta^{(B)}(\omega_{P_1}) + \beta^{(A)}(\omega_S) - \beta^{(B)}(\omega_{P_2}) - \beta^{(A)}(\omega_I), \quad \text{for BS} \tag{1.23}$$
$$\Delta\beta = \beta^{(B)}(\omega_{P_1}) + \beta^{(A)}(\omega_S) - \beta^{(A)}(\omega_{P_2}) - \beta^{(B)}(\omega_I),$$

and

$$\Delta\beta = \beta^{(A)}(\omega_{P_1}) - \beta^{(A)}(\omega_S) + \beta^{(A)}(\omega_{P_2}) - \beta^{(A)}(\omega_I),$$
$$\Delta\beta = \beta^{(B)}(\omega_{P_1}) - \beta^{(B)}(\omega_S) + \beta^{(A)}(\omega_{P_2}) - \beta^{(A)}(\omega_I), \quad \text{for OPC} \tag{1.24}$$
$$\Delta\beta = \beta^{(B)}(\omega_{P_1}) - \beta^{(A)}(\omega_S) + \beta^{(A)}(\omega_{P_2}) - \beta^{(B)}(\omega_I).$$

A common measure of signal Conversion Efficiency (CE) is the ratio of the idler power at the waveguide output (waveguide length L_{wg}) to the signal input power:

$$\text{CE} = \frac{P_I(z = L_{\text{wg}})}{P_S(z = 0)}. \tag{1.25}$$

For given input powers P_{P_1}, P_{P_2}, and P_S, with variable modes m, o, p, q, of the interacting waves, the idler power can be approximated by [11]:

$$P_{\rm I}(L_{\rm wg}) = 4 \cdot \left(\gamma^{(mopq)}\right)^2 \cdot L_{\rm wg}^2 \cdot |\eta_{\rm FWM}(\Delta\beta)|^2 \cdot P_{\rm P_1} P_{\rm P_2} P_{\rm S} \cdot {\rm e}^{-\alpha \cdot L_{\rm wg}}, \tag{1.26}$$

where α is the waveguide attenuation. The nonlinearity coefficient is defined as

$$\gamma^{(mopq)} = \frac{\omega_0 n_2}{c_0 A_{\rm eff}^{(mopq)}} \tag{1.27}$$

with the nonlinear refractive index n_2 of the waveguide material. The FWM efficiency $\eta_{\rm FWM}(\Delta\beta)$ is determined by

$$\eta_{\rm FWM}(\Delta\beta) = \frac{1 - {\rm e}^{-(\alpha + {\rm j}\Delta\beta)L_{\rm wg}}}{(\alpha + {\rm j}\Delta\beta)L_{\rm wg}}. \tag{1.28}$$

The cross effective area $A_{\rm eff}$ between the involved modes is given by [10]

$$A_{\rm eff}^{(mopq)} = \frac{\sqrt{I^{(m)} I^{(o)} I^{(p)} I^{(q)}}}{\left| \iint \left[\left(\vec{\Psi}^{(m)}\right)^* \cdot \vec{\Psi}^{(o)} \right] \left[\left(\vec{\Psi}^{(p)}\right)^* \cdot \vec{\Psi}^{(q)} \right] {\rm d}A \right|} \tag{1.29}$$

with $I^{(X)} = \iint |\vec{\Psi}^{(X)}|^2 {\rm d}A$, where ${\rm d}A$ denotes integration over the waveguide cross section. It can be shown that the maximum FWM efficiency of Eq. (1.28) is independent of the waveguide attenuation α and is achieved when $\Delta\beta = 0$, which is the primary objective in waveguide optimization. However, one additional degree of freedom is given by the selection of the interacting modes, which determines the cross effective area Eq. (1.29). For specific mode combinations, the denominator is very small, which substantially reduces conversion efficiency. Figure 1.7 depicts the intensities for two configurations of modes of a typical (regular) waveguide. The surface plot visualizes the integrand of the denominator of Eq. (1.29), the black solid lines show the cross sections along $y = 100$ nm. It is obvious that the integral over Fig. 1.7a will give a relatively large positive value, while the integral over Fig. 1.7b will be close to zero, since positive and negative lobes cancel each other. Thus, the first case results in a much higher idler power.

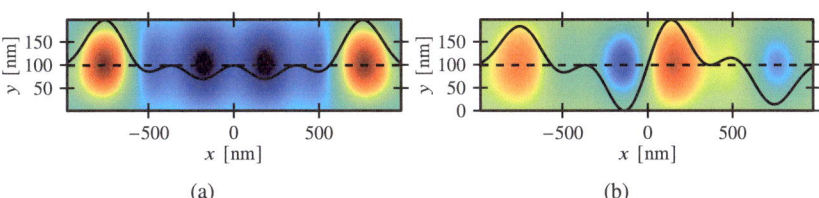

Fig. 1.7 a Good nonlinear interaction with $(m, o, p, q) = ({\rm TE}_3, {\rm TE}_1, {\rm TE}_2, {\rm TE}_2)$. **b** Poor nonlinear interaction with $(m, o, p, q) = ({\rm TE}_3, {\rm TE}_0, {\rm TE}_2, {\rm TE}_2)$.

We discussed earlier that good PM can be only achieved in realistic waveguides if the interacting modes are selected according to Eq. (1.23) or Eq. (1.24), i.e., if either one mode is selected for all lasers, or if two modes are used in specific arrangements for two lasers, respectively. These selections yield favorable overlap integrals according to Fig. 1.7a, characterized by positive lobes and high integral magnitudes.

1.2.3 Optimization of the SOI-Waveguide Design

The target of the performed optimization is to maximize the bandwidth of optical wavelength conversion within or between dedicated optical bands. In this respect, we put the focus on the combinations of signal and idler wavelengths depicted in Fig. 1.8.

Our goal is achieved by optimizing the waveguide geometry, i.e., rib width, slab height, number of dips, dip width, and dip depth [6, 12]. The range of the varied parameters was adjusted with the waveguide manufacturer IHP and are subject to practicability. The propagating modes were selected from the set $\{TE_0, TE_1, TE_2\}$, and the combinations were restricted by the PM conditions Eq. (1.23) for BS and Eq. (1.24) for OPC. The optimizer could either choose freely from one-mode and two-mode configurations, or it was forced to use two-mode configurations (labeled *2M* in Table 1.1). For every realization, the normalized FWM efficiency

$$\left| \frac{\eta_{FWM}(\Delta\beta)}{\max|\eta_{FWM}|^2} \right| \qquad (1.30)$$

was evaluated for a 2 cm long waveguide, yielding values strictly confined to the range [0, 1]. The pump waves were always placed arbitrarily in the O-, S-, C-, and L-bands. The FWM bandwidth is defined as the frequency range over which the signal can be varied, while keeping the pump wavelengths fixed, such that the FWM efficiency remains above −3 dB (i.e., at least 50% of its peak value). In this evaluation, a signal-to-pump separation of at least 5 nm was enforced. Table 1.1 lists the main results.

In the following, **three realizations** of Table 1.1, reflecting three ways to achieve PM, are outlined in detail:

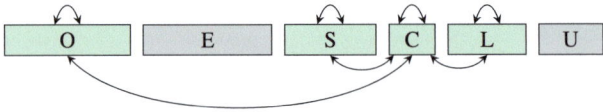

Fig. 1.8 Signal and idler combinations covered in this project. O, E, S, C, L, and U are the low-attenuation wavelength-bands of optical fibers

1 Ultra-Broadband Photonic Signal-Processor

Table 1.1 Optimization results of our work [6, 12]. Each scenario defines the nonlinear wavelength-conversion process (BS or OPC followed by *source band → target band*), and whether two-mode operation was enforced (*2M*) or the optimizer was allowed to use either one or two modes. The used modes are aligned in the order: P_1, S, P_2, I. The last column shows the obtained/maximum FWM bandwidth B_{FWM} for the respective realization, where the maximum depends on the width of source and target bands, the FWM process, the pump placement, and the enforced signal to pump separation of 5 nm

Scenario	Modes [TE_x]	w_{rib} [nm]	h_{slab} [nm]	Num. Dips	w_{dip} [nm]	h_{dip} [nm]	B_{FWM} [nm]
BS, O→O	0, 1, 1, 0	3000	70	2	400	150	52/95
BS, S→S	2, 2, 2, 2	1500	70	0	200	70	65/65
BS, C→C	2, 2, 2, 2	1500	70	0	200	70	29/30
BS, L→L	2, 2, 2, 2	1500	70	0	200	70	54/55
BS, C→O	1, 2, 2, 1	1500	70	1	300	70	35/35
BS, C→S	2, 2, 2, 2	1500	70	0	200	70	35/35
BS, C→L	2, 2, 2, 2	1500	70	0	200	70	35/35
BS, O→O, 2M	0, 1, 1, 0	3000	70	2	400	150	52/95
BS, S→S, 2M	1, 2, 1, 2	2000	70	2	200	150	61/65
BS, C→C, 2M	2, 1, 2, 1	1800	70	1	200	150	29/30
BS, L→L, 2M	2, 1, 2, 1	1600	70	1	200	120	53/55
BS, C→O, 2M	1, 2, 2, 1	1500	70	1	300	70	35/35
BS, C→S, 2M	2, 1, 1, 2	1650	100	1	400	70	35/35
BS, C→L, 2M	2, 1, 1, 2	2100	100	1	400	70	35/35
OPC, C→C	2, 2, 2, 2	1750	70	2	200	70	35/35
OPC, C→C, 2M	2, 1, 1, 2	1800	100	1	400	70	19/30

BS, C → O: Corresponds to line 5 of Table 1.1. This scenario is called *classical case*, since it is also applied in experiments in optical fibers [11, 13]. S and P_2 are placed in the TE_2 mode, P_1 and I are in the TE_1 mode. The group delay curves of this realization are shown in Fig. 1.9.

The curves are approximately parallel. Arranging modes and wavelengths as outlined, the group delay at the mean wavelength of the TE$_1$ waves matches the group delay at the mean wavelength of the TE$_2$ waves, i.e.,

$$\tau_{gr}^{TE_1}\left(\frac{\omega_{P_1}+\omega_I}{2}\right) = \tau_{gr}^{TE_2}\left(\frac{\omega_{P_2}+\omega_S}{2}\right). \tag{1.31}$$

Condition Eq. (1.31) is equivalent to full PM. This condition is preserved if the signal is shifted towards longer wavelengths, as the idler will follow in the case of BS, and the group delays at mean wavelengths will stay close to each other over a wide wavelength range. Figure 1.10a shows the normalized FWM efficiency for the waveguide geometry and P$_1$ wavelength of this realization. It can be concluded that maximum conversion efficiency for C → O band conversion can be achieved over the entire C-band. This behavior exhibits a high robustness if rib width and slab height are varied around their optimum values. One can see in Fig. 1.10b that manufacturing imperfections can be tolerated with this scenario, since the FWM bandwidth remains large over a wide range.

Since the production of SOI waveguides was restricted by the etching process, this optimized solution could not be realized, as further boundary conditions had to be observed: we had to fix the SOI height to 220 nm, slab height to 100 nm and could only optimize the rib width. Furthermore it was problematic to etch dips. The optimized alternative solution for the BS, C → O scenario that has been identified did not require dips, and used TE$_0$ and TE$_1$ modes for the interacting waves in the arrangement 0, 1, 1, 0 according to the definition in Table 1.1. It has a rib width of 1672 nm, and the required slab height of 100 nm and SOI height of 220 nm. The device likewise shows the classical case of PM and exhibits the full FWM bandwidth of 35 nm. This solution was then produced and characterized; results are presented in Sect. 1.4.1.

BS, C → S: Corresponds to line 6 of Table 1.1. It can be seen in Fig. 1.11 that here the TE$_2$ mode has a very low dispersion in the S- and C-band. Thus, placing all waves in the TE$_2$ mode, PM and, in consequence, full conversion efficiency is achieved over a very wide wavelength range due to the low dispersion, see Fig. 1.12a. The sensitivity with respect to deviations of rib width and slab height from their optimum values is likewise low, as can be seen in Fig. 1.12b. Due to these features, this realization is extremely versatile and applicable to various scenarios.

In Table 1.1 the applications BS, C → L, and OPC, C → C likewise make use of the low dispersion of the TE$_2$ mode in the S-, C-, and L-band.

1 Ultra-Broadband Photonic Signal-Processor

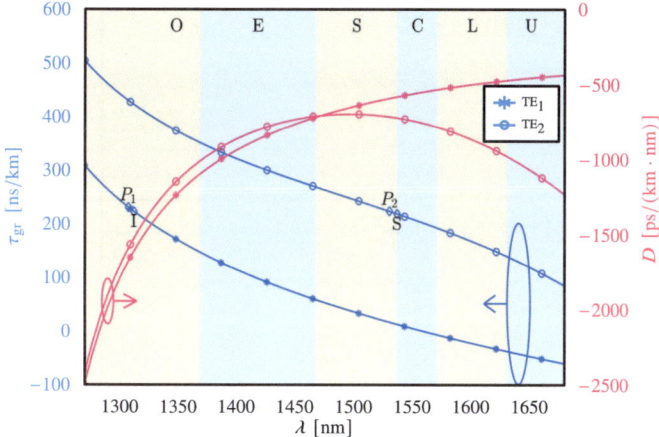

Fig. 1.9 Group delay and dispersion of the realization BS, C → O scenario [12]

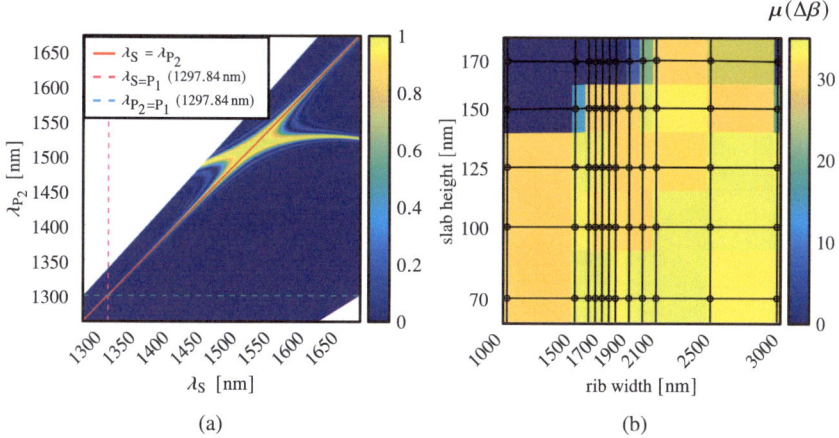

Fig. 1.10 a Four-Wave Mixing efficiency for BS, C → O scenario. **b** Four-Wave Mixing bandwidth μ [nm] for BS, C → O scenario under variations of rib width and slab height of the SOI waveguide [12]

BS, S → S: Corresponds to line 9 of Table 1.1. Here, both pump lasers are placed in the TE_1 mode at very short wavelengths, signal and idler in the S-band use the TE_2 mode. When the signal wavelength is varied, the average group delay of the pumps remains constant, as both pump wavelengths are unchanged. Due to the low dispersion of the TE_2 mode (see Fig. 1.13), the group delay at the mean signal and idler wavelength is likewise almost constant, and Eq. (1.31) holds again. However, this process is very sensitive to the placement of the pump wavelengths. The flat (yellow) region of optimum FWM efficiency in

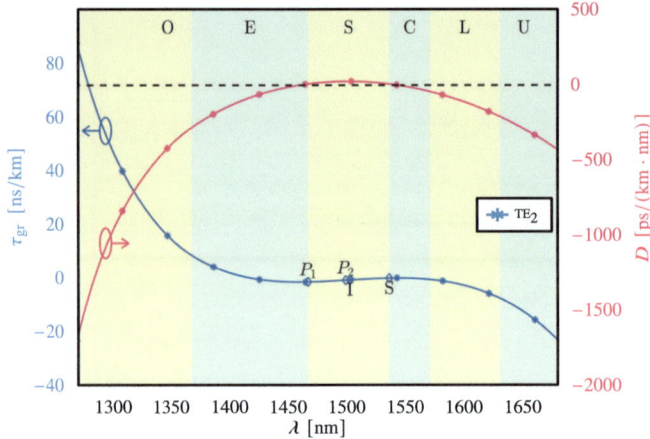

Fig. 1.11 Group delay and dispersion of the realization BS, C → S scenario [12]

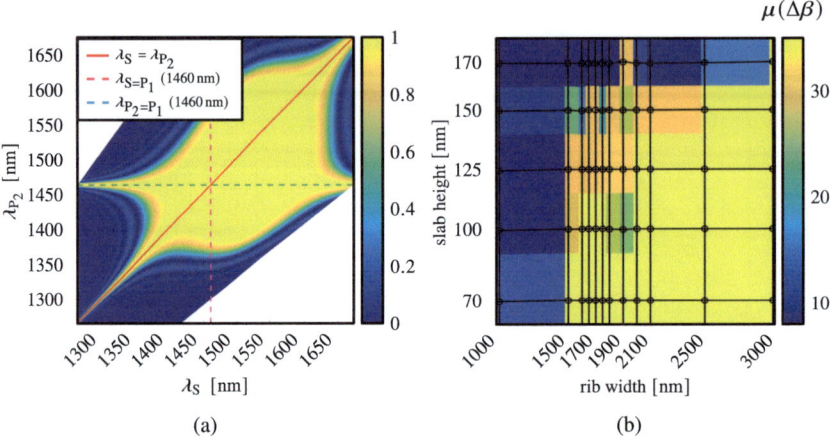

Fig. 1.12 a Four-Wave Mixing efficiency for BS, C → S scenario. **b** Four-Wave Mixing bandwidth μ [nm] for BS, C → S scenario under variations of rib width and slab height of the SOI waveguide [12]

Fig. 1.14a indicates that the wavelength of P_2 (and likewise that of P_1) should be very stable to maintain PM. The process also shows a high sensitivity with respect to variations of rib width and slab height, as this realization requires an exact tuning of the group delay curves, which depend on the waveguide geometry (see Fig. 1.14b).

Dips are a very effective means to fine tune the group delay curves of selected modes, and are therefore frequently used in the optimization of the scenarios in Table 1.1. Figure 1.15

1 Ultra-Broadband Photonic Signal-Processor

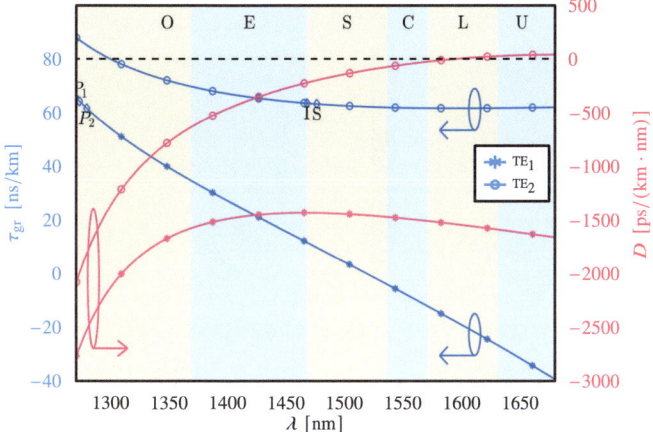

Fig. 1.13 Group delay and dispersion of the realization BS, S → S scenario [12]

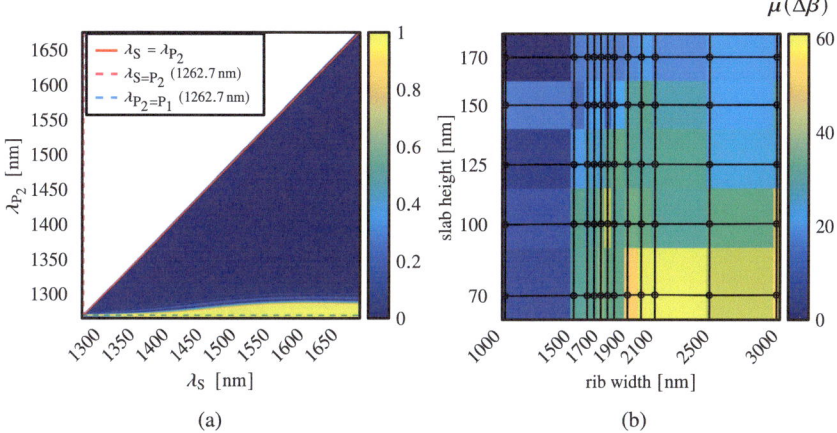

Fig. 1.14 a Four-Wave Mixing efficiency for BS, S → S scenario. **b** Four-Wave Mixing bandwidth μ [nm] for BS, S → S scenario under variations of rib width and slab height of the SOI waveguide [12]

shows three-dimensional evaluations of the FWM efficiency-bandwidth of the three discussed scenarios with respect to dip parameters. The size of the balls has the same information as the color, i.e., the larger the ball, the higher the bandwidth. For each ball, rib width and slab height have been set to their optimum values, respectively. As can be derived from the figure, the first two scenarios are rather insensitive to the imprinted number of dips, including no dip at all. The last FWM scenario, however, requires specific dip parameters for a sufficiently large FWM bandwidth.

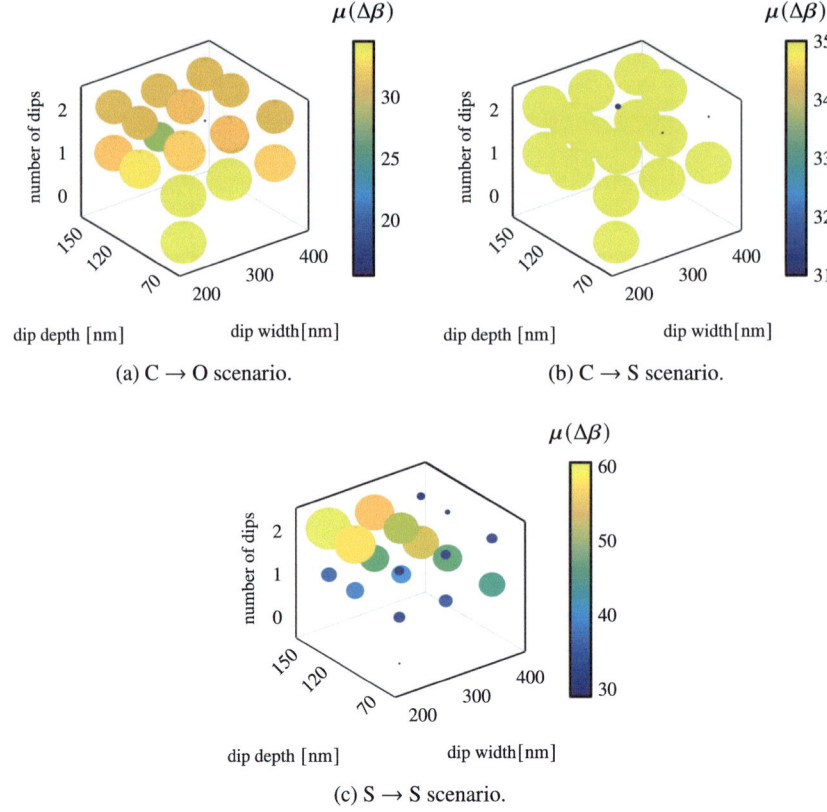

Fig. 1.15 Four-Wave Mixing bandwidth w.r.t. dip parameters for BS [12]

1.2.4 Modeling Nonlinear Wave Propagation in SOI Waveguides

Calculating the idler power at the end of the SOI waveguide with Eq. (1.26) is an approximation, which is well applicable for the optimization of the waveguide design. However, it does not cover several relevant effects. To be able to predict the idler power along the waveguide with high accuracy, we developed a Continuous Wave (CW) frequency domain propagation simulation framework, which computes complex amplitudes as the waves propagate along the waveguide. The effects of mode dependent attenuation, linear mode coupling and nonlinear coupling are all considered here.

We model the nonlinear polarization vector as

$$\vec{P}^{\mathrm{nl}} = \iiint \epsilon_0 \overset{\leftrightarrow}{\chi}^{[3]}(\tau_\zeta, \tau_\eta, \tau_\rho) \vdots \vec{E}(t-\tau_\zeta)\vec{E}(t-\tau_\eta)\vec{E}(t-\tau_\rho) \, \mathrm{d}\tau_\zeta \, \mathrm{d}\tau_\eta \, \mathrm{d}\tau_\rho, \quad (1.32)$$

where $\overset{\leftrightarrow}{\chi}^{[3]} \in \mathbb{C}^{3\times3\times3\times3}$ is the third-order nonlinear susceptibility tensor, and $\vdots$ denotes the tensor product. The expression in component (sum) notation reads

$$P_i^{\mathrm{nl}} = \iiint \sum_{j,k,\ell} \epsilon_0 \chi^{[3]}_{ijk\ell}(\tau_\zeta, \tau_\eta, \tau_\rho) \cdot E_j(t-\tau_\zeta) E_k(t-\tau_\eta) E_\ell(t-\tau_\rho) \, \mathrm{d}\tau_\zeta \, \mathrm{d}\tau_\eta \, \mathrm{d}\tau_\rho \quad (1.33)$$

with indices $i, j, k, \ell \in \{x, y, z\}$. We consider only third-order nonlinearities, as even-order nonlinear effects are absent in centrosymmetric materials such as silicon due to inversion symmetry [14]. Furthermore, higher-order odd nonlinearities (e.g., fifth-order and beyond) are typically significantly weaker and can therefore be neglected. In the following, we adopt the common assumption that the perturbation vector $\vec{P}'$ in Eq. (1.13) introduces a z-dependence in the propagation amplitudes $\hat{E}^{(m)}_{\omega_a}$, but does not influence the transversal mode profile $\vec{\Psi}^{(m)}_{\omega_a}$. Since the outer gradient Ⓐ in Eq. (1.13) alters the transversal field distribution, this assumption is only valid if $\vec{P}'$ does not contribute to these variations. This is ensured by requiring $\mathrm{div}\, \vec{P}' = 0$. Under this condition, evaluating Eq. (1.13) at the positive carrier frequency ω_0 yields

$$\Delta \underline{\vec{E}}_{\omega_0} + \mathrm{grad}\left(\frac{1}{\epsilon_\mathrm{r}'} \cdot \underline{\vec{E}}_{\omega_0} \cdot \mathrm{grad}\,\epsilon_\mathrm{r}'\right) - \mu_0 \cdot \epsilon_0 \cdot \epsilon_\mathrm{r}' \cdot \partial_t^2 \underline{\vec{E}}_{\omega_0} = \mu_0 \cdot \partial_t^2 \underline{\vec{P}}'_{\omega_0}. \quad (1.34)$$

After some algebraic manipulation and calculus, and under the assumption $\partial_z \epsilon_\mathrm{r}' = 0$, the x-component of Eq. (1.34) can be stated as

$$\sum_m \Bigg[\underbrace{\left\{\left(\partial_x^2 + \partial_y^2\right)\Psi^{(m)}_{x,\omega_0} + \left(\epsilon_\mathrm{r}'\beta_0^2 - \beta^{2(m)}_{\omega_0}\right)\Psi^{(m)}_{x,\omega_0} \partial_x\left[\frac{1}{\epsilon_\mathrm{r}'}\left(\Psi^{(m)}_{x,\omega_0}\partial_x \epsilon_\mathrm{r}' + \Psi^{(m)}_{y,\omega_0}\partial_y \epsilon_\mathrm{r}'\right)\right]\right\}}_{Ⓑ}$$

$$\cdot \hat{E}^{(m)}_{\omega_0} e^{\mathrm{j}\left(\omega_0 t - \beta^{(m)}_{\omega_0} z\right)} + \left(\partial_z^2 \hat{E}^{(m)}_{\omega_0} - 2\mathrm{j}\beta^{(m)}_{\omega_0} \partial_z \hat{E}^{(m)}_{\omega_0}\right) \Psi^{(m)}_{x,\omega_0} e^{\mathrm{j}\left(\omega_0 t - \beta^{(m)}_{\omega_0} z\right)} \Bigg]$$

$$= \mu_0 \partial_t^2 \underline{\vec{P}}'_{x,\omega_0} \quad (1.35)$$

with $\beta_0 = \omega_0/c_0$. Since the field amplitude $\hat{E}^{(m)}_{\omega_0}$ has only a z dependence due to the perturbation $\vec{P}'$, all terms involving $\partial_z \hat{E}^{(m)}_{\omega_0}$ vanish in the homogeneous case ($\vec{P}' = \vec{0}$). Mode orthogonality then implies that each summand must vanish individually, i.e., $\mathcal{B} = 0, \forall m$. Assuming $\vec{P}'$ does not affect the mode profiles $\vec{\Psi}_{\omega_0}^{(m)}$, this condition also holds for $\vec{P}' \neq \vec{0}$. Extending Eq. (1.35) to all vector components leads to

$$\sum_m -j\beta_{\omega_0}^{(m)} \partial_z\left(\hat{E}_{\omega_0}^{(m)}\right) \cdot \vec{\Psi}_{\omega_0}^{(m)} \, e^{j\left(\omega_0 t - \beta_{\omega_0}^{(m)} z\right)} = \frac{1}{2}\mu_0 \partial_t^2 \vec{P}'_{\omega_0}, \tag{1.36}$$

where the slowly varying envelope approximation is used, i.e., $\left|\partial_z^2 \hat{E}_{\omega_0}^{(m)}\right| \ll \left|2\beta_{\omega_0}^{(m)} \partial_z \hat{E}_{\omega_0}^{(m)}\right|$ and $\left|\partial_z \hat{E}_{\omega_0}^{(m)}\right| \ll \left|\beta_{\omega_0}^{(m)} \hat{E}_{\omega_0}^{(m)}\right|$. The nonlinear component of $\vec{P}'$ arises from the third-order susceptibility tensor $\overset{\leftrightarrow}{\chi}^{[3]}$, which governs interactions among three optical fields to produce a nonlinear response at frequency ω_0. By inserting Eqs. (1.15) into (1.32) and retaining only frequency combinations $(\omega_\zeta, \omega_\eta, \omega_\rho)$ containing two positive and one negative frequency, we account for relevant nonlinear processes such as BS and OPC. The corresponding nonlinear material polarization at frequency ω_0 is then given by

$$\vec{P}_{\omega_0}^{\mathrm{nl}} = \sum_{\substack{(\zeta\eta\rho)\\ \in S}} \sum_{\substack{(m_1 m_2 m_3)\\ \in M}} \frac{3}{4}\epsilon_0 \hat{E}_{\omega_\zeta}^{(m_1)} \hat{E}_{\omega_\eta}^{(m_2)} \left(\hat{E}_{\omega_\rho}^{(m_3)}\right)^*$$

$$\cdot \overset{\leftrightarrow}{X}^{[3]}(\omega_\zeta, \omega_\eta, \omega_\rho) : \vec{\Psi}_{\omega_\zeta}^{(m_1)} \vec{\Psi}_{\omega_\eta}^{(m_2)} \left(\vec{\Psi}_{\omega_\rho}^{(m_3)}\right)^* e^{j(\omega_0 t - \delta\beta z)},$$

$$S = \left\{(\zeta\eta\rho) : \omega_0 = \omega_\zeta + \omega_\eta + \omega_\rho \,\big|\, \omega_{\zeta,\eta} > 0, \omega_\rho < 0\right\},$$

$$M = \left\{(m_1 m_2 m_3) \,\big|\, m_{1,2,3} \in \{\mathrm{TE}_0, \mathrm{TE}_1, \ldots\}\right\},$$

$$\delta\beta = \beta_{\omega_\zeta}^{(m_1)} + \beta_{\omega_\eta}^{(m_2)} - \beta_{\omega_\rho}^{(m_3)}, \tag{1.37}$$

where $\overset{\leftrightarrow}{X}^{[3]}$ is the Fourier transform of $\overset{\leftrightarrow}{\chi}^{[3]}$, i.e., $\mathcal{F}\left\{\overset{\leftrightarrow}{\chi}^{[3]}\right\} = \overset{\leftrightarrow}{X}^{[3]}$. By inserting Eqs. (1.37) into (1.36), multiplying from the left by $\left(\vec{\Psi}_{\omega_0}^{(a)}\right)^*$, and integrating over the transverse plane (exploiting mode orthogonality) we derive the propagation equation for mode (a) at frequency ω_0 corresponding to [15]

$$\partial_z \hat{E}_{\omega_0}^{(a)} = -\underbrace{\frac{\alpha^{(a)}}{2} \hat{E}_{\omega_0}^{(a)}}_{\text{attenuation}} - j\tilde{\gamma} \sum_m \underbrace{C_{(m)}^{(a)}}_{\text{mode coup.}} \hat{E}_{\omega_0}^{(m)} e^{-j\Delta\beta_{\mathrm{lin}} z} \tag{1.38}$$

$$- j\frac{3\tilde{\gamma}}{4} \sum_{\substack{(\zeta\eta\rho)\\ \in S}} \sum_{\substack{(m_1 m_2 m_3)\\ \in M}} \underbrace{N_{(m_1 m_2 m_3)}^{(a)}}_{\text{nonlin. coefficient}} \cdot \hat{E}_{\omega_\zeta}^{(m_1)} \hat{E}_{\omega_\eta}^{(m_2)} \left(\hat{E}_{\omega_\rho}^{(m_3)}\right)^* e^{-j\Delta\beta z},$$

where

$$\Delta\beta_\mathrm{lin} = \beta^{(m)}_{\omega_0} - \beta^{(a)}_{\omega_0},$$

$$\Delta\beta = \beta^{(m_1)}_{\omega_\zeta} + \beta^{(m_2)}_{\omega_\eta} - \beta^{(m_3)}_{\omega_\rho} - \beta^{(a)}_{\omega_0},$$

$$C^{(a)}_{(m)} = \iint \left(\vec{\Psi}^{(a)}_{\omega_0}\right)^* \delta\epsilon_r \vec{\Psi}^{(m)}_{\omega_0} dA,$$

$$N^{(a)}_{(m_1 m_2 m_3)} = \iint \left(\vec{\Psi}^{(a)}_{\omega_0}\right)^* \cdot \left(\overset{\leftrightarrow}{X}^{[3]}(\omega_\zeta, \omega_\eta, \omega_\rho) : \vec{\Psi}^{(m_1)}_{\omega_\zeta} \vec{\Psi}^{(m_2)}_{\omega_\eta} \left(\vec{\Psi}^{(m_3)}_{\omega_\rho}\right)^*\right) dA.$$

$$\tilde{\gamma} = \frac{\beta_0^2}{2\beta^{(a)}_{\omega_0} \iint \left|\vec{\Psi}^{(a)}_{\omega_0}\right|^2 dA},$$

$$\alpha^{(a)} = \frac{\beta_0^2 \epsilon''_r}{\beta^{(a)}_{\omega_0}},$$

The linear coupling coefficient $C^{(a)}_{(m)}$ accounts for interactions between modes at the same frequency, whereas the nonlinear coefficient $N^{(a)}_{(m_1 m_2 m_3)}$ describes mode coupling across all frequency combinations. The third-order susceptibility tensor $\overset{\leftrightarrow}{X}^{[3]}$ mediates these nonlinear processes by characterizing the material's response to combinations of interacting optical fields. The computation of $\overset{\leftrightarrow}{X}^{[3]}$ in silicon under specific symmetry constraints is summarized in [15, 16]. These coupling mechanisms give rise to a system of coupled differential equations of the form given in Eq. (1.38), which must be solved numerically. The efficiency of both linear and nonlinear processes is governed not only by energy conservation and the associated phase mismatches $\Delta\beta_\mathrm{lin}$ and $\Delta\beta$, but also by the spatial overlap of the interacting mode fields.

1.2.5 Numerical Evaluation

In our simulation, we couple three optical waves (two pumps and one signal) at distinct, discrete frequencies into an SOI waveguide. Linear mode coupling enables each wave to excite all available modes at its respective frequency. In addition, nonlinear interactions generate new frequency components in all supported modes, resulting from all possible combinations of the input frequencies. The efficiency and characteristics of both linear and nonlinear interactions depend on the phase mismatches, denoted by $\Delta\beta_\mathrm{lin}$ and $\Delta\beta$, as well as on the corresponding coupling coefficients, $C^{(a)}_{(m)}$ and $N^{(a)}_{(m_1 m_2 m_3)}$. As a result, the number of potential interactions increases rapidly with the number of modes and frequencies involved. To manage this complexity, we restrict the simulation to light propagating only in the three lowest-order guided modes of the waveguide: TE_0, TE_1, and TE_2. Additionally, we only consider frequency components that are expected to be generated with non-negligible efficiency. Specifically, we simulate the evolution of the input frequencies ω_{P_1}, ω_{P_2}, ω_S, as well as the two generated frequencies ω_{BS}, ω_{OPC} which arise from BS and OPC FWM processes. Despite this simplification, the system still involves solving 15 coupled differential equations, as detailed in Table 1.2.

Table 1.2 Coupled differential equations to be solved simultaneously

	ω_{P_1}	ω_{P_2}	ω_S	ω_{OPC}	ω_{BS}
TE$_0$	$\partial_z \hat{E}^{(TE_0)}_{\omega_{P_1}}$	$\partial_z \hat{E}^{(TE_0)}_{\omega_{P_2}}$	$\partial_z \hat{E}^{(TE_0)}_{\omega_S}$	$\partial_z \hat{E}^{(TE_0)}_{\omega_{OPC}}$	$\partial_z \hat{E}^{(TE_0)}_{\omega_{BS}}$
TE$_1$	$\partial_z \hat{E}^{(TE_1)}_{\omega_{P_1}}$	$\partial_z \hat{E}^{(TE_1)}_{\omega_{P_2}}$	$\partial_z \hat{E}^{(TE_1)}_{\omega_S}$	$\partial_z \hat{E}^{(TE_1)}_{\omega_{OPC}}$	$\partial_z \hat{E}^{(TE_1)}_{\omega_{BS}}$
TE$_2$	$\partial_z \hat{E}^{(TE_2)}_{\omega_{P_1}}$	$\partial_z \hat{E}^{(TE_2)}_{\omega_{P_2}}$	$\partial_z \hat{E}^{(TE_2)}_{\omega_S}$	$\partial_z \hat{E}^{(TE_2)}_{\omega_{OPC}}$	$\partial_z \hat{E}^{(TE_2)}_{\omega_{BS}}$

Once all relevant third-order susceptibilities $\overset{\leftrightarrow}{\chi}^{[3]}(\omega_\zeta, \omega_\eta, \omega_\rho)$ are determined for the frequencies involved, the nonlinear coupling coefficients $N^{(a)}_{(m_1 m_2 m_3)}$ can be computed accordingly. The linear coupling coefficients $C^{(a)}_{(m)}$ are governed by the coupling matrix $\delta\epsilon_r$, which models perturbations in the refractive index profile due to fabrication imperfections, surface roughness, or mechanical stress. Because $\delta\epsilon_r$ depends on various uncertain parameters, its exact analytical form is often unknown. Consequently, it is common to use a heuristic approach to adjust $\delta\epsilon_r$ such that simulated linear coupling matches experimental results. The resulting system of coupled-mode equations Eq. (1.38) is integrated numerically using a variable-order, variable-step Adams–Bashforth–Moulton solver, taking into account both linear and nonlinear effects.

An SOI waveguide is considered, featuring a rib width of 1800 nm, a slab height of 100 nm, a total silicon height of 220 nm, and a propagation length of 2 cm. A single dip structure with a width of 400 nm and a depth of 70 nm is included. The waveguide geometry, illustrated in Fig. 1.16, is identical to that described in [12].

Table 1.3 summarizes the selected input and output wavelengths, their associated modes, and the input power levels. These values are chosen to optimize the efficiency of the OPC process. Realistic attenuation coefficients are included for each mode.

The resulting power evolution along the waveguide without linear mode coupling ($C^{(a)}_{(m)} = 0$) is depicted in Fig. 1.17a. The propagation of the input waves (—■—, —✳—,

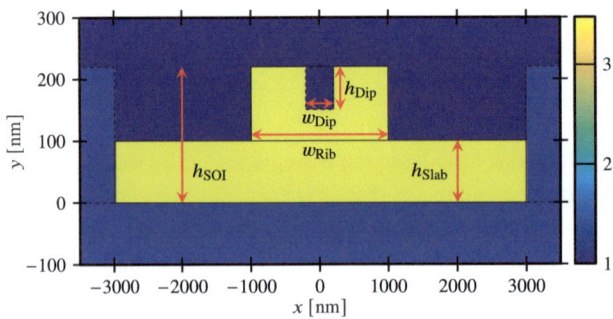

Fig. 1.16 Refractive index geometry of the NR waveguide, indicating slab, SOI, and dip heights, along with rib and dip widths [15]

Table 1.3 Simulation parameters: Wavelengths and modes of propagating waves with high conversion efficiency, input powers, and mode-dependent attenuation coefficients. The asterisk on I_{OPC^*} indicates that PM was optimized for the OPC process

	Input [nm]			Generated [nm]		Input Power [dBm]			α [dB/cm]
	P_1	P_2	S	I_{OPC^*}	I_{BS}	P_1	P_2	S	
TE_0	–	–	–	–	–	–	–	–	1
TE_1	–	1535	–	1519	1529	–	20	–	2
TE_2	1524	–	1540	–	–	20	–	10	3

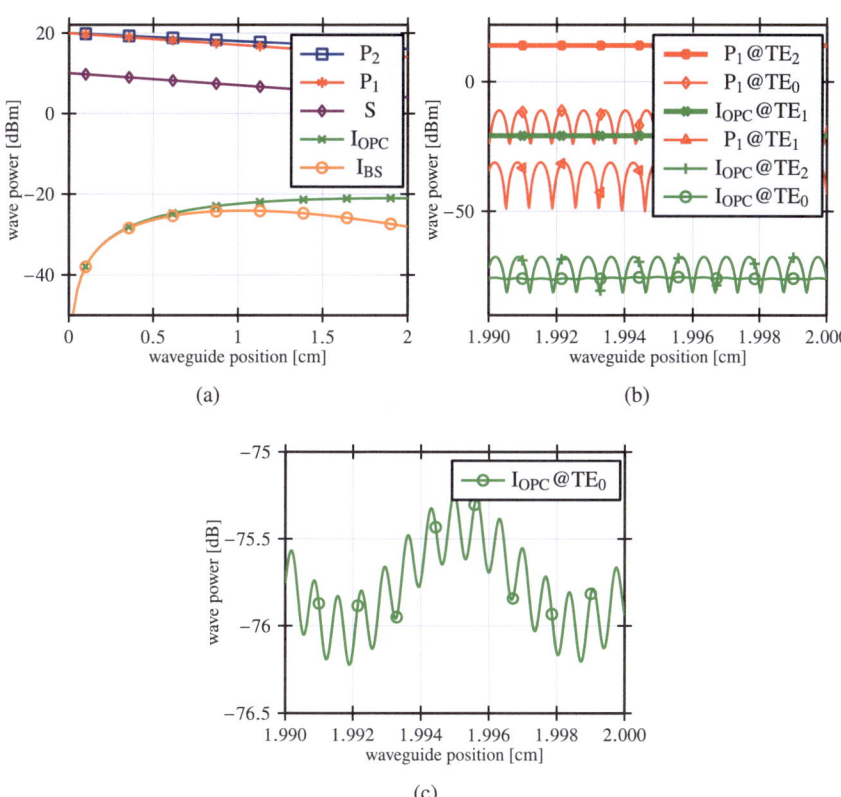

Fig. 1.17 **a** Simulated power evolution of pumps, signal, and idlers along the waveguide without linear mode coupling. The mode distribution corresponds to Table 1.3. **b** Simulated power evolution of selected waves including both linear and nonlinear coupling. **c** Vertical zoom on the OPC idler in TE_0 mode from (**b**) [15]

and ──⊷──) is primarily influenced by intrinsic linear waveguide attenuation. Nonlinear depletion due to energy transfer to the OPC and BS idlers is only marginally visible in the logarithmic scale. Among the generated idlers, the OPC signal (──✶──) exhibits the highest conversion efficiency due to optimal PM, while the BS idler (──⊙──) shows reduced efficiency as a result of a larger nonlinear phase mismatch $\Delta\beta$. All other potential FWM products involving different frequency-mode combinations are highly phase-mismatched and therefore negligible in power. To investigate the impact of linear mode coupling, we heuristically define the refractive index perturbation matrix as

$$\delta\epsilon_r = \begin{bmatrix} 1 & 0 & 0 \\ 0 & 0 & 0 \\ 0 & 0 & 0 \end{bmatrix},$$

which introduces mode coupling by breaking the orthogonality between guided modes in the calculation of $C^{(a)}_{(m)}$. This simplification neglects the minor but relevant y and z components of the transverse mode fields. As emphasized, a more realistic form of $\delta\epsilon_r$ should ultimately be determined by matching simulations with experimental results. Figure 1.17b illustrates the power evolution in the presence of additional linear mode coupling. In principle, all waves can couple into all supported modes, but we exemplarily highlight one pump and one idler wave (───, ───). The oscillation periods of the coupled powers are governed by the respective phase mismatches $\Delta\beta_{\text{lin}}$ according to $L_{\text{Oscillation}} = \frac{2\pi}{|\Delta\beta_{\text{lin}}|}$, while the amount of coupled power depends on both $C^{(a)}_{(m)}$ and $\Delta\beta_{\text{lin}}$. For example, the normalized linear coupling coefficient of the OPC idler from mode TE_1 to TE_0 is

$$\frac{\left| C^{(TE_0)}_{(TE_1)} \right|}{\iint \left| \vec{\Psi}^{(TE_0)}_{\omega_{\text{IOPC}}} \right|^2 dA} = 1.3 \times 10^{-4} \quad (\text{──⊙──}). \tag{1.39}$$

In contrast, coupling from TE_1 to TE_2 yields

$$\frac{\left| C^{(TE_2)}_{(TE_1)} \right|}{\iint \left| \vec{\Psi}^{(TE_2)}_{\omega_{\text{IOPC}}} \right|^2 dA} = 3.6 \times 10^{-3} \quad (\text{──✚──}). \tag{1.40}$$

The corresponding phase mismatches are

$$\begin{aligned} |\beta^{(TE_1)}_{\omega_{\text{IOPC}}} - \beta^{(TE_0)}_{\omega_{\text{IOPC}}}| &= 9.1 \times 10^4 \, \text{m}^{-1} \quad (\text{──⊙──}), \\ |\beta^{(TE_1)}_{\omega_{\text{IOPC}}} - \beta^{(TE_2)}_{\omega_{\text{IOPC}}}| &= 9.2 \times 10^5 \, \text{m}^{-1} \quad (\text{──✚──}). \end{aligned} \tag{1.41}$$

Although the coupling coefficient from TE_1 to TE_2 is higher (which would generally lead to stronger coupling), its significantly larger phase mismatch reduces the effective transfer of power. As a result, both TE_0 and TE_2 exhibit similar power levels for the OPC idler (—⊖—, —+—). Figure 1.17c presents a zoomed-in view of I_{OPC} in mode TE_0. Two superimposed oscillations are visible: a slow oscillation over the entire propagation length due to direct coupling from TE_1, and a faster, weaker oscillation resulting from second-order coupling via TE_2.

1.3 Design and Fabrication of All-Optical Signal Processors and Their Components

Nonlinear optical signal processing in the silicon material system is severely hampered by effects associated with free carriers that are inevitably created due to Two-Photon Absorption (TPA). Free carriers introduce loss and therefore pose a challenge to nonlinear devices using CW light in the telecom and datacom wavelength range. However, by studying free carrier lifetime in silicon nano-waveguides [17] we could show early on that PIN assisted waveguide structures very effectively suppress free carrier associated loss and therefore permit e.g. much enhanced FWM efficiency in singlemode waveguides [18]. The here presented work for the first time extends the approach of using PIN assisted silicon waveguides for nonlinear optical signal processing to the domain of multimode waveguides. The sub-structures of the here-studied photonic signal processor were fabricated in IHP's 200 mm BiCMOS pilotline cleanroom facilities using SG25_PIC or SG25H5_EPIC technology, contrasting with earlier experiments that relied on dedicated technological flows [19]. Our photonic integrated circuits have, thus, been manufactured using a full technology flow which at the same time served a multi-project wafer client base. The advantage of this approach compared to other techniques that increase nonlinear efficiency is the use of standard SOI-material, which is fully compatible with additional thermal budgets typically encountered at later stages of device fabrication or integration. In addition these advanced technologies provide much enhanced system integration perspectives. This chapter will provide a brief overview of the photonic integrated circuits technology platform used for the fabrication of multimode PIN assisted waveguide structures followed by CW device characterization.

1.3.1 Foundry Design

In recent years silicon photonic design flows have aligned more and more with application specific electronic circuit design because of progress of design tools as well as due to maturing of process design kits. Today silicon photonic foundry users typically expect the support of electronic design automation on several tools. The full design flow applicable to the designs fabricated in SG25_PIC or SG25H5_EPIC is shown in Fig. 1.18. The depicted principal design flow assumes the availability of all photonic building blocks in the design kit libraries. However, this was not the case for the integrated multimode optics that was

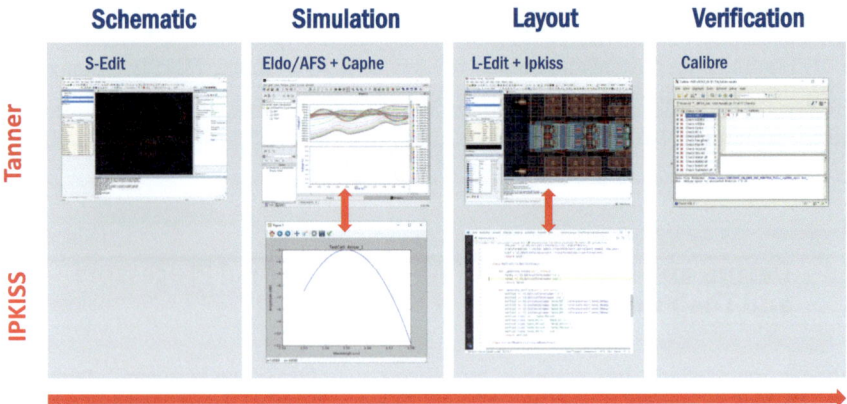

Fig. 1.18 Principal design flow for photonic integrated circuits and electronic photonic integrated circuits in IHP's silicon photonic technologies. The photonic integrated circuits realized for the present study were mostly designed using the IPKISS framework of Luceda Photonics

designed independently as described in the previous chapter plus additional optimization using mode propagation software. Layout and physical verification made use of the fab supported software.

1.3.2 E/O Waveguide Platform

IHP's electronic photonic integrated circuit technology was first demonstrated in 2013 [20] and was technologically matured and optimized considerably over the course of several years. The most recent technology generation features for example >60 GHz germanium photodiodes and fast hetero-junction bipolar transistors [21]. SG25_PIC and SG25H5_EPIC are matched to feature virtually identical silicon photonics. A sketch of a cross-section of the photonic integrated circuit technology is depicted in Fig. 1.19. The integrated structures were fabricated on 220 nm SOI substrates. The etching depth of the grating was 70 nm. Then the waveguides were etched using a nitride hardmask. The E/O waveguide was designed as a rib-waveguide with PIN junction embedded. For waveguide fabrication 248 nm Deep-UV scanner lithography was deployed. Two implantation steps with high doping concentrations ($\approx 1 \times 10^{20}$ cm^{-3}) defined the contact area for completion of the PIN waveguide diode process. Then, following salicidation W-contact plugs were realized on top of high-dose implanted areas to allow for connection to metal pads. The contact plugs have the shape of inverted cones. The backend of line consisted of a 5 metal layer interconnect corresponding to standard Metal 1 (Al) up to Top-metal 2 separated by inter-layer dielectrics. The wafer was passivated with silicon nitride.

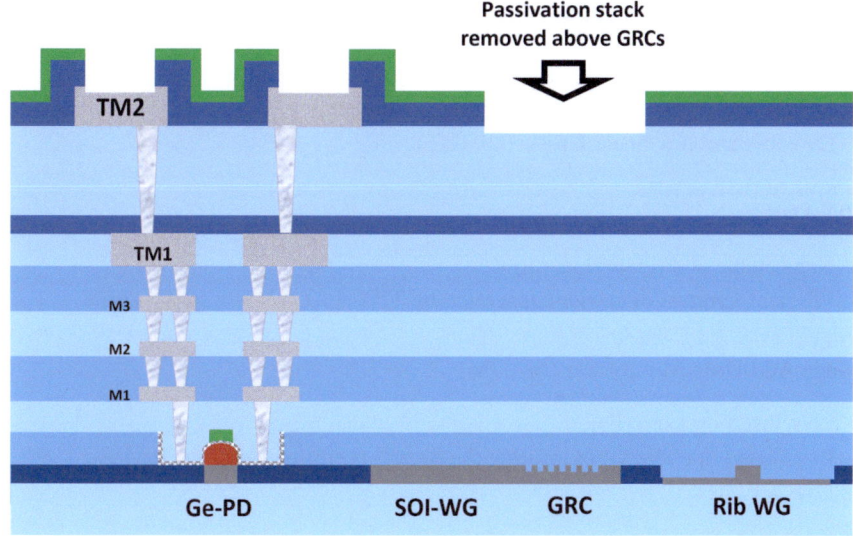

Fig. 1.19 Exemplary cross section of IHP's photonic integrated circuit technology SG25_PIC. WG = waveguide, GRC = grating coupler, PD = photodiode, TM = top-metal, M = metal

1.3.3 Components and Integration Into an All-Optical Signal Processor

To transfer the results of the theory in Sect. 1.2 into practical devices, the fabrication constraints outlined before must be considered. Therefore, typical devices will require multimode bends to keep the footprint within chip boundaries, a PIN-diode integrated along the waveguide to reduce TPA-induced Free Carrier Absorption (FCA) at high pump powers, individual Mode Add-Drop Multiplexers (MADMs) to excite the proper higher-order modes within the MMWG and Grating Couplers (GCs) adjusted for the individual wavelengths of pumps, signals and idlers. The collection of these components is illustrated in Fig. 1.20. The design goals for those components can be summarized as follows.

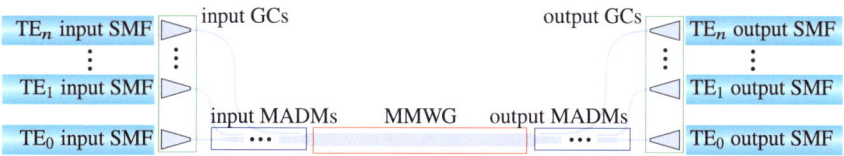

Fig. 1.20 Sketch of a typical AOWC with its main components. The waveguide is drawn straight, but can be wrapped into a spiral to reduce the footprint

- Multi-mode Waveguide (MMWG) and Bends

  - Optimized dispersion for ultra-broadband operation in different bands.
  - Small effective area for large nonlinear coefficient.
  - Low loss and low linear Cross-Talk (XT).

- PIN-Diode

  - Neglectable additional linear loss.
  - Efficient removal of carriers generated by TPA.

- Mode Add-Drop Multiplexer (MADM)

  - Low Insertion Loss (IL).
  - Broadband operation, not limiting the overall performance of the devices.
  - Low XT.

- Grating Coupler (GC)

  - Low IL across the relevant bands for pump, signal, and idler.

In the following, design considerations for individual components are discussed.

1.3.3.1 Multi-mode Waveguide and Bend Design

The overall PM optimization was extensively discussed in Sect. 1.2. In practical devices however, the tuning parameters are more limited. Therefore, no dips were used for most devices, and in all but one device, the standard slab thickness $h_{\text{slab}} = 100$ nm was used. In the following paragraph a MMWG width of $w_{\text{rib}} = 1200$ nm and wavelength of $\lambda = 1550$ nm is assumed, when not stated otherwise.

In order to allow compact devices, multi-mode bends have been designed by using linearly-graded curvature *Euler* bends [22]. While for the same footprint (measured by the effective radius R_{eff}) the maximum curvature is larger than for a circular bend, the mode mismatch at the straight-to-bent interfaces is reduced. This mode mismatch is typically the dominant XT mechanism for circular multi-mode bends. Both bend types are depicted in Fig. 1.21.

The resulting transmission and XT of TE_1 excitation for $R = R_{eff} = 100$ μm are calculated and shown in Fig. 1.22.

The XT of 21.9 dB achieved for the circular bend may seem sufficient, but for a compact design, a large number of bends is required and the XT may add up coherently to large powers. For example, even a small design with 8 of those circular bends would result in unacceptable 21.9 dB $- 20\log_{10}(8) = 3.8$ dB worst-case XT. The simulated Euler bends

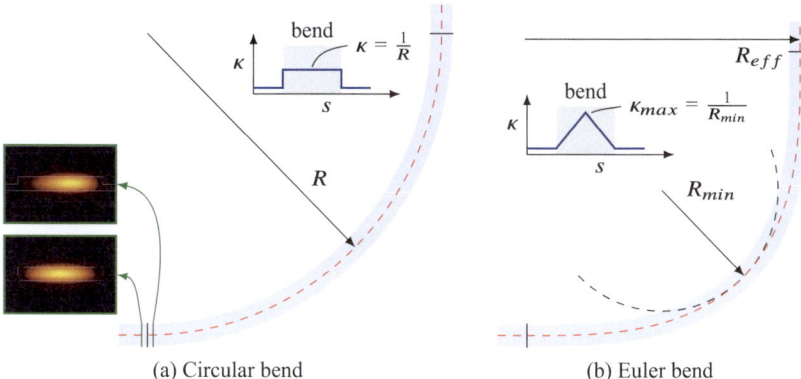

Fig. 1.21 Schematic of the two bend types considered in this work: **a** *circular* bends with piecewise constant curvature κ, which exhibit a mode mismatch at the straight-to-bent interface (both mode fields as green-bounded insets); and *Euler* bends with piecewise linear curvature. The red dashed lines indicate waveguide center paths, which are the reference for all radii and curvature

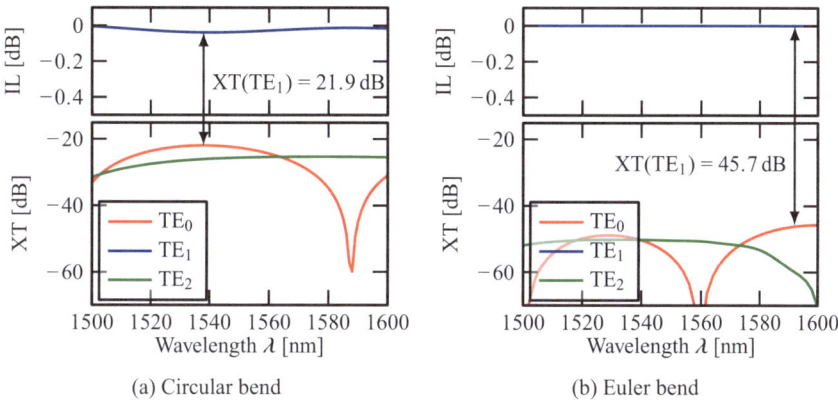

Fig. 1.22 Transmission of MMWG bends with $w_{\text{rib}} = 1.2\,\mu\text{m}$, $h_{\text{slab}} = 0.1\,\mu\text{m}$ excited by 1550 nm in the TE_1 mode. Arrows indicate worst ratio of IL and XT

on the other hand, exhibit a high XT supression of 45.7 dB per bend, which yields enough headroom for larger designs. For actual devices Euler bends with R_{eff} of 50 μm and 100 μm have been used, where the larger radius was used in later devices to further suppress XT.

1.3.3.2 Design of Mode Add-Drop Multiplexer

As all external signals are single-moded when coupled onto the Photonic Integrated Circuit (PIC), on-chip MADMs are mandatory to excite the higher-order modes within the MMWGs. Although various implementations of on-chip MADMs have been proposed,

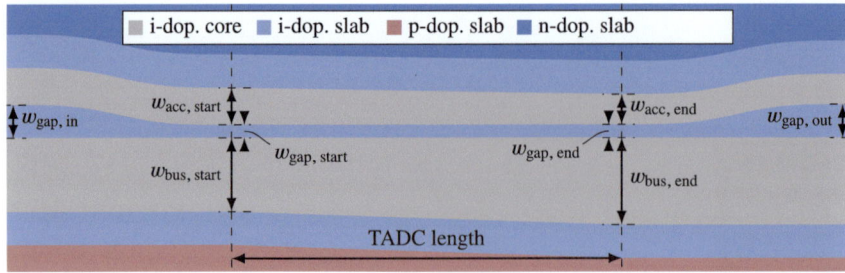

Fig. 1.23 Schematic top view of a TADC used to implement mode multiplexing and its relevant dimensions

Tapered Asymmetric Directional Couplers (TADCs) [23] seem to be among the most popular choices and have been used in this work. They consist of a larger (multi-mode) *bus* waveguide and a smaller (single-mode) *access* waveguide which are brought close together. When both waveguides are close, one of them is tapered to stimulate an adiabatic power exchange between the access waveguide and a higher-order mode of the bus waveguide. The structure is depicted in Fig. 1.23. Each cross section can be described by the widths of the bus and access waveguide (w_{bus} and w_{acc}) and the gap between them (w_{gap}). In order to achieve the desired mode conversion, the effective index n_{eff} of the access waveguide's fundamental mode must be close to n_{eff} of the mode in the bus waveguide that should be coupled. This n_{eff}-matching is shown in Fig. 1.24. Points of mode coupling are marked with dashed circles, and the continuous lines must be followed slowly through these points to achieve adiabatic mode conversion.

For the example case of a TE_1-MADM, Fig. 1.25 shows the transmission, when the access waveguide is linearly tapered from 480 nm to 530 nm while the bus waveguide width is fixed at 1190 nm. In principle both widths can be tapered individually or simultaneously. With IL less then 0.2 dB and XT better than 40 dB in the entire C-band, the simulated component is close to ideal.

1.3.3.3 Design of PIN Diode to Suppress Free-Carrier Absorption

The usage of a reverse-biased PIN diode has been explored only for single-mode waveguides so far [17]. The extension towards multi-mode waveguides is one main contribution of this work. Figure 1.26 gives an overview of the different doping regions that form the PIN diode. The intrinsic region is introduced where the optical mode is guided to prevent additional linear optical losses in the doped regions. Moderately doped *p*- and *n*-regions are added next to the intrinsic region. These have lower optical losses than the heavily doped *n*+- and *p*+-regions next to the contacts. There is a clear trade-off between better electrical performance (faster carrier removal at lower voltages) and better optical performance (lower linear losses), which can be optimized in the design phase.

1 Ultra-Broadband Photonic Signal-Processor

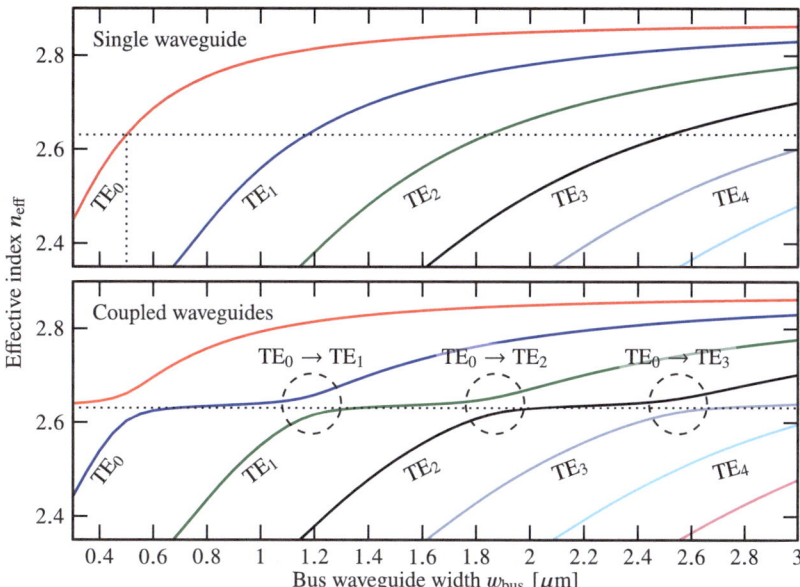

Fig. 1.24 Effective index n_{eff} for isolated (top) and coupled (bottom) waveguides with $h_{\text{slab}} = 100$ nm, $w_{\text{gap}} = 200$ nm and $w_{\text{acc}} = 500$ nm at $\lambda = 1550$ nm

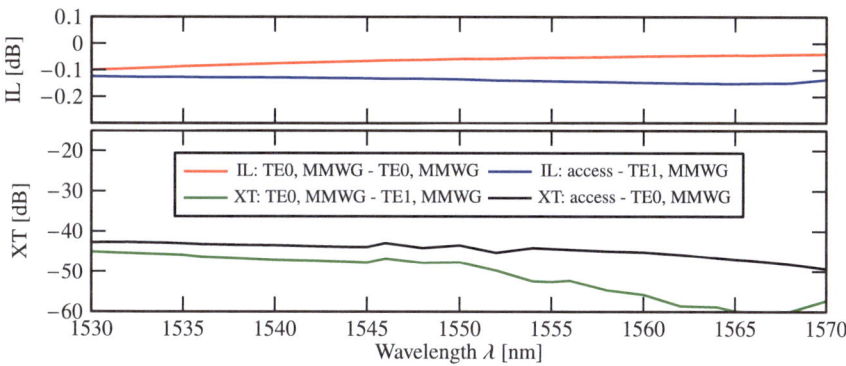

Fig. 1.25 Simulated transmission for TE_1-MADM designed for $\lambda = 1550$ nm with $w_{\text{gap}} = 200$ nm, $w_{\text{bus}} = 1190$ nm and w_{acc} linearly tapered from 480 nm to 530 nm in the TADC section

When the geometry parameters w_{rib} and h_{slab} are already used to optimize the FWM phase matching, and h_{SOI} and the doping levels are given by the PIC/Electronic-Photonic Integrated Circuits (EPIC) process, only the separation between waveguide edge and doping region d_{dop} can be optimized during design stage. The widths of the p- and n-doping regions had virtually no influence on the diode performance, as long as it is large enough such that the high-doping regions do not overlap with the mode.

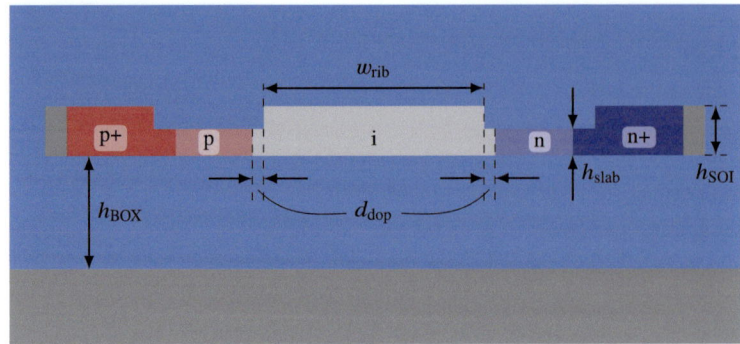

Fig. 1.26 Cross section and dimensions of MMWG with integrated PIN diode. Red, white and blue shades indicate different levels of p-, i- and n-doping

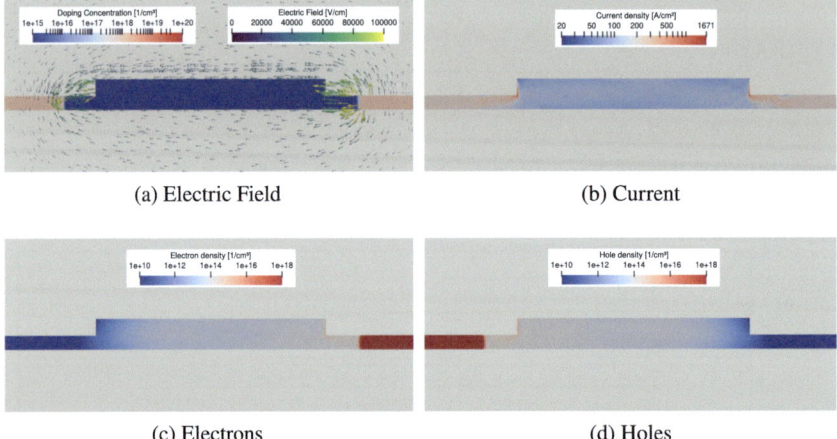

Fig. 1.27 Steady-state solution of reverse-biased PIN diode under optical excitation

The steady-state carrier densities, currents and fields were calculated using a drift-diffusion model based on DEVSIM [24]. The standard model was extended by Auger and impact-ionization processes to properly model high-field effects. The TPA carrier generation rate is calculated from the imaginary part of the nonlinear coefficient in Eq. (1.38). The carrier distribution was then used to derive the FCA losses by an overlap with the modes. A typical steady-state solution is shown in Fig. 1.27. When this procedure is repeated with $w_{\text{rib}} = 1.7\,\mu\text{m}$, $d_{\text{dop}} = 0.25\,\mu\text{m}$, an optical input power of 100 mW respectively 500 mW is coupled to the TE_0 mode at a wavelength of 1550 nm, an I-V-characteristic as in Fig. 1.28 is obtained.

The current shows a typical diode behavior. Above about 1 V the forward current starts to increase exponentially (which is accompanied by carrier injection and very high optical

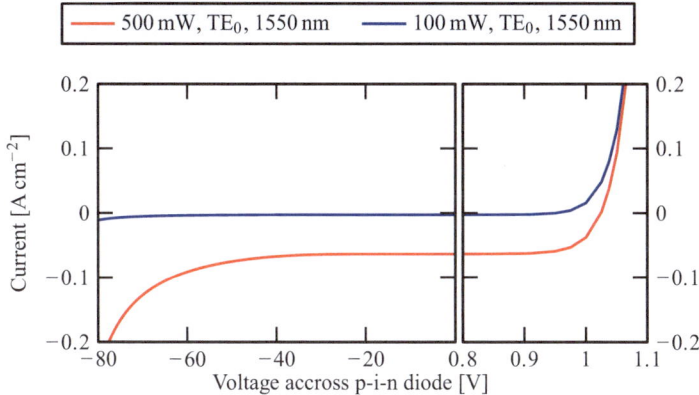

Fig. 1.28 Electrical I-V-characteristic of an MMWG with $w_{rib} = 1.7\,\mu m$, $h_{slab} = 0.1\,\mu m$, $d_{dop} = 0.25\,\mu m$ excited by 100 mW (red) and 500 mW (blue) in the TE_0 mode

loss). With large negative bias the diode will breakdown with a breakdown voltage that depends on the injected optical power. In between is the desired operation region, where current is almost constant with voltage and carriers are efficiently removed at saturation speed.

As a next step, the optical performance was optimized by varying d_{dop}. For this purpose, $w_{rib} = 1.7\,\mu m$ is still assumed, while now two pumps are injected, 100 mW at 1310 nm into TE_0 and 100 mW at 1550 nm into TE_1. This resembles the conditions used in the following system experiments of Sect. 1.4. The results for selected bias voltages are shown in Fig. 1.29. Here, larger reverse bias voltages lead to reduced losses as expected, however with diminishing returns beyond -10 V. Due to the larger intrinsic region for large d_{dop}, the field strength at low voltages is not sufficient to saturate the current in the intrinsic region anymore. Therefore there is an optimum d_{dop}. For 0 V this is around $0.1\,\mu m$, for larger bias voltages the optimum barely exists and $d_{dop} \geq 0.4\,\mu m$ is typically sufficient for low optical losses. In the following, $d_{dop} = 0.5\,\mu m$ was selected for all devices, as this value yielded good overall performance across many bias and optical input conditions.

1.3.4 Summary of Fabricated Devices

During this work, a variety of devices has been fabricated with different target applications in mind. Table 1.4 summarizes these devices with their respective parameters.

The devices marked with an asterisk are further investigated in system experiments in Sect. 1.4. For devices with dips, the loss was too high to perform further characterizations. In generations IV and V no separate MADM test structures have been implemented. Therefore, losses are only estimated by assuming a 1 dB loss of each MADM.

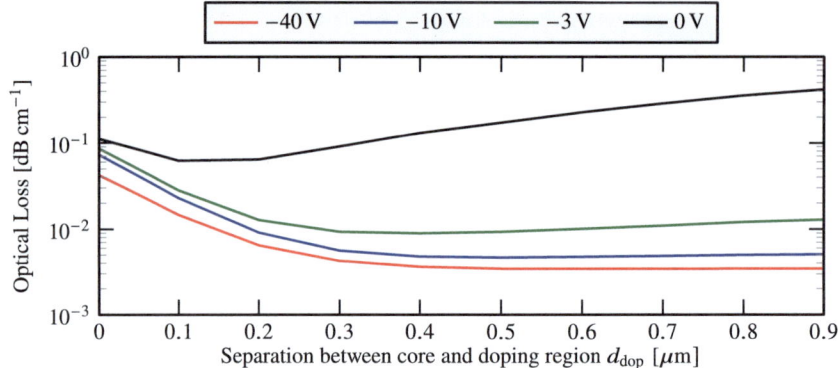

Fig. 1.29 Optical losses of the TE_0 mode at $\lambda = 1550$ nm in a reverse-biased-diode MMWG with $w_{rib} = 1.7$ μm and $h_{slab} = 100$ nm under different bias conditions. Optical excitation is by 100 mW at 1310 nm in TE_0 and 100 mW at 1550 nm in TE_1

1.3.5 Component Characterization

Particularly, in the first device generation several test structures have been implemented to characterize performance of individual components. After fabrication of each iteration those structures have been characterized to improve the following designs.

1.3.5.1 Mode Add-Drop Multiplexer

In order to evaluate the MADM performance, the first-generation device with almost-zero MMWG length is used, where IL and XT from the MMWG is neglectable. By measuring all input-output combinations, the graphs in Fig. 1.30 were generated. As the normalized transmission through two MADMs is better than −2 dB, the IL for each MADM is better than 1 dB over the complete C-band. At the same time, XT suppression is better than 20 dB. This does not fully achieve the predicted values of Fig. 1.25, but is well usable in the envisaged applications.

1.3.5.2 Multi-mode Waveguide Dispersion and Loss

As the waveguide group index and dispersion are *the* critical parameters to achieve phase matching, in a first step group index was measured by using a Photonic Dispersion and Loss Analyzer (PDLA). The group delays through the 3 paths $TE_{0,in} \rightarrow TE_{0,out}$, $TE_{1,in} \rightarrow TE_{1,out}$ and $TE_{2,in} \rightarrow TE_{2,out}$ were measured for different MMWG lengths. By a linear fit the constant group delay of coupling fibers and MADMs was separated from the MMWG group delay and group index of each mode was calculated. Looking at the results in Fig. 1.31, the measurements resemble the predictions from simulation quite well. For the

Table 1.4 Summary of fabricated devices, their design parameters and measured performance

Design					Characterization			
Gen.	Width [nm]/Features	Lengths [mm]	FWM process	Mode	GCs' center wavelength [nm]	Worst dip for 11.7 mm [dB]	Loss [dB/cm]	
I	1190	7.1, 11.7, 22.3	C to L WLC	TE_0	1550	2.6	0.8	
				TE_1	1550	6.9	2.5	
				TE_2	1550	21.1	7.0	
I	1850	7.1, 11.7, 22.3	C to L WLC	TE_0	1550	1.7	0.8	
				TE_1	1550	5.5	1.8	
				TE_2	1550	5.2	1.8	
II*	1672	7.1, 11.7	C to O WLC	TE_0	1310	3.2	0.9	
				TE_1	1550	13.8	1.8	
II	1698	7.1, 11.7	C to O WLC	TE_0	1310	2.7	0.9	
				TE_1	1550	16.9	1.8	
II	1722	7.1, 11.7	C to O WLC	TE_0	1310	3.8	0.9	
				TE_1	1550	15.8	1.8	
III	1640/with dips, larger bends	27.1	C to O WLC	TE_0	1310	–	High	
				TE_1	1550	–	High	
III	1850/larger bends	38.1	C to O WLC	TE_0	1310	–	High	
				TE_1	1550	–	High	
III	1800/with dips, larger bends	7.2, 16.9, 27.3, 38.3	C to O WLC	TE_0	1310	–	High	
				TE_1	1550	–	High	
IV	1800/with dips, larger bends	15.9	C to C OPC	TE_1	1550	–	High	
				TE_2	1550	–	High	
IV	1700/straight, $h_{slab}=150$ nm	12.1	C to L WLC	TE_1	1520	< 1	~2.5	
				TE_2	1580	1	~1	
IV	1700/straight, $h_{slab}=150$ nm	10.7	C to S WLC	TE_1	1490	1	~2	
				TE_2	1520	1.5	~4	
IV*	1750/straight	8.3	C to S OPC	TE_2	1520	< 1	~4	
V	1700 / $h_{slab}=72$ nm, larger bends	30.0	multipurpose S/C/L	TE_2	1550	> 35	~2	

second-generation device there is a slight vertical offset, but this does not shift the phase matched wavelengths.

With the same measurements, IL and XT were also determined. As can be seen in Fig. 1.32, the IL of the first-generation devices is mostly flat across the spectrum with few deeper dips at certain wavelengths. Looking at the XT traces at the same wavelengths, it is clear that this dips originate from strong XT carrying a substantial amount of power into

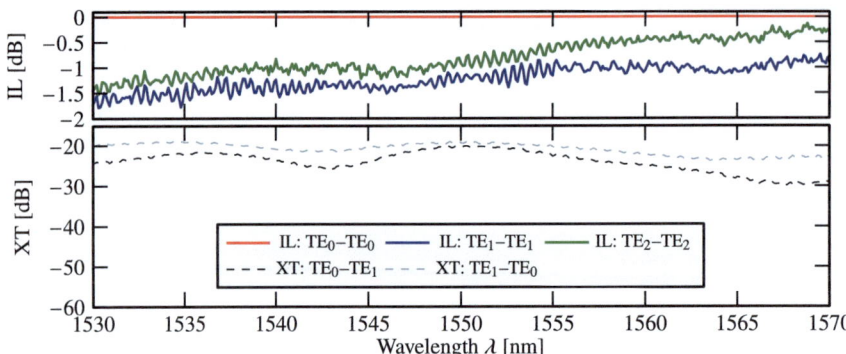

Fig. 1.30 Measured IL and XT for first-generation device with two back-to-back MADMs (only showing two dominant XT contributors). All traces are normalized to the TE_0–TE_0 case to cancel effects of GCs

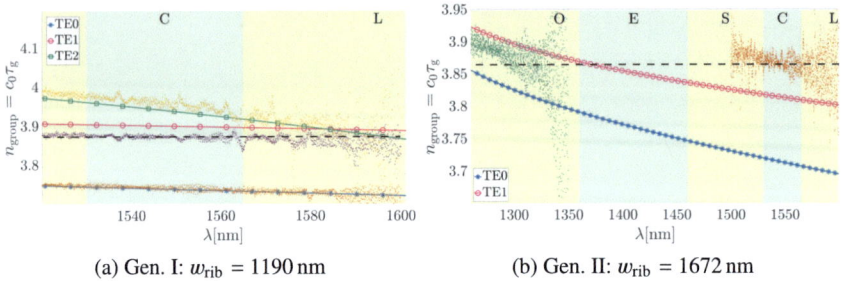

(a) Gen. I: $w_{rib} = 1190$ nm

(b) Gen. II: $w_{rib} = 1672$ nm

Fig. 1.31 Group index simulations (solid) and measurements (dots) of the first three (two) guided modes for: **a** first and **b** second generation devices. Black dashed lines indicate phase matching

other modes. The magnitudes of those dips and the IL for different modes are collected in Table 1.4. The IL spectra for the second-generation MMWGs are given in Fig. 1.33. As the GCs at the TE_0 (TE_1) ports are limited to the O-band (C-band), the XT can not be determined directly for these devices. Though the high XT measured in the C-band did not severely degrade the conversion efficiency in the C- to O-band experiments (see Fig. 1.36a and additional results in [25]), our aim was to reduce this linear crosstalk between the waveguide modes in successive generations.

In the fourth generation, to overcome the high XT observed so far, MMWGs have also been implemented as simple straight waveguide. Due to limited chip sizes, the largest MMWG length was 12.1 mm. This approach could reduce the ripples in the IL trace down to about 1 dB and increase the XT suppression up to about 20 dB, which can be seen in Fig. 1.34. This is a substantial improvement compared to the results in Figs. 1.32 and 1.33. This straight waveguide is further investigated in the system experiments in the following sections.

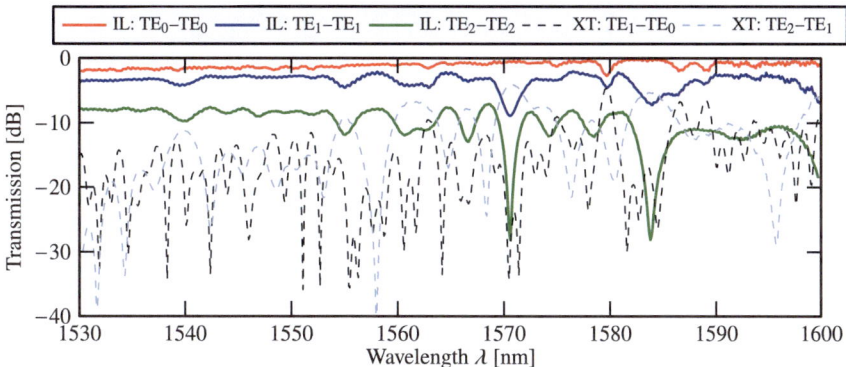

Fig. 1.32 Measured IL and XT spectra for a $w_{rib} = 1190$ nm MMWG with 11.7 mm length (only showing two strongest XT contributors). All transmissions are normalized to back-to-back GC response

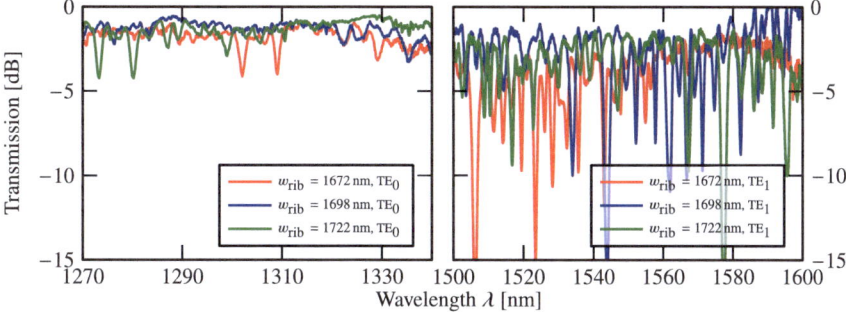

Fig. 1.33 Second generation MMWGs designed for C- to O-band conversion: Measured transmission of 11.7 mm long devices normalized to $TE_0 - TE_0$ back-to-back responses for TE_0 (left) and TE_1 (right)

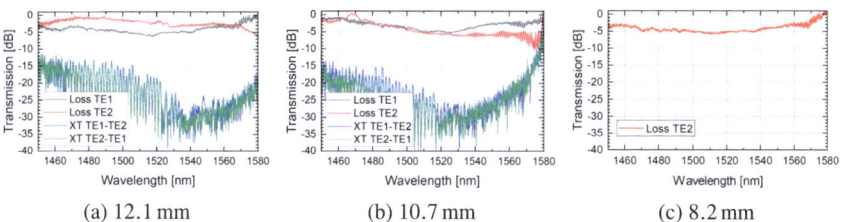

(a) 12.1 mm (b) 10.7 mm (c) 8.2 mm

Fig. 1.34 Measured transmission of fourth-generation devices with lengths of **a** 12.1 mm, **b** 10.7 mm and **c** 8.3 mm as described in Table 1.4

1.4 Performance Evaluation of SOI Waveguides for Ultra-Broadband All-Optical Wavelength Converters and Optical Phase Conjugation

In this section, we present the experimental verification and system-level evaluation of silicon photonic chips designed for ultra-broadband wavelength conversion and Optical Phase Conjugation (OPC). Building upon the theoretical principles, dispersion engineering, and design methodologies outlined in the previous chapters, we focus on the key performance metrics, the experimental setup used for testing, and the interpretation of results in both linear and nonlinear regimes. The objective is to validate the chip performance under real-world signal conditions and assess its suitability for deployment in long-haul and high-capacity optical networks.

The PICs were characterized using a temperature-stabilized coupling stage maintained at a constant temperature by a temperature controller to ensure stable operating conditions. Since all the PICs under test used GCs for light coupling, the PICs were connected using two cleaved SSMFs, which could be adjusted for arbitrary coupling angles at the PIC input and output. Two six-axis manual positioning stages were employed to achieve efficient coupling, providing full control over both linear (x, y, z) and spherical (r, θ, ϕ) coordinates. For the characterization and system measurements, the input and output coupling angles were matched and tuned to achieve symmetric transmission around the desired center wavelength. As the PICs discussed in this chapter all utilize GCs, which are inherently polarization-dependent, a Polarization Controller (PC) was placed before the input to adjust the state of polarization and ensure optimal alignment with the grating coupler's acceptance profile.

As it was discussed in Sect. 1.3.5, linear characterization—including insertion loss and bandwidth—was performed to establish a baseline reference. With the linear properties characterized, we proceeded to the nonlinear evaluation of the PICs to gain insights into the behavior and limitations of the FWM process under varying optical power levels. Nonlinear characterization was carried out using a CW optical signal and pump, both polarized to match the TE mode of the waveguide. They were combined using a WDM or a 3-dB coupler. The combined output, containing both signal and pump, was coupled into the PIC via a grating coupler. The idler generated through the FWM process was then measured at the output using an Optical Spectrum Analyzer (OSA). By sweeping the input signal power, the FWM efficiency and idler generation could be evaluated. The signal CE was defined as the ratio of the idler power at the waveguide output to the signal power at the waveguide input, as shown in Eq. (1.25). These nonlinear measurements help to define the optimal operating conditions for the PICs and establish a robust foundation for system-level integration and performance optimization.

After completing both linear and nonlinear characterizations, we advanced to the system-level experiments. In this phase, a data signal was generated at the transmitter and received by the receiver. To establish a reference for the Optical Signal-to-Noise Ratio (OSNR) penalty introduced by the transmission system, a noise-loading stage was implemented. In this setup, the Bit Error Rate (BER) was recorded at various OSNR levels and compared against the theoretical BER curve derived from the Additive White Gaussian Noise (AWGN) model. With the OSNR penalty characterized, full system measurements were conducted using high-data-rate signals, either in single-channel or WDM configurations, employing advanced modulation formats.

In the following sections, we discuss the nonlinear characterization results, as well as full system measurements, performed on two devices: a multi-mode SOI waveguide designed for ultra-broadband C $\rightarrow$ O wavelength conversion, and a multi-mode SOI straight waveguide designed for ultra-broadband C $\rightarrow$ C OPC.

1.4.1 Multi-mode SOI Waveguide for Ultra-Broadband C $\rightarrow$ O Wavelength Conversion

1.4.1.1 Experimental Setup

The PIC employed for ultra-broadband C $\rightarrow$ O wavelength conversion in this section was designed as described in Sect. 1.2.3. Figure 1.35 illustrates the setup used for the following measurements [26]. It consists of the transmitter, the AOWC stage, the transmission stage and finally a Kramers-Kronig Receiver (KK-Rx). The transmitter was used to generate a 32 GBd Single-Polarization (SP) Quadrature Phase-Shift Keying (QPSK) signal. In the AOWC, this signal was combined with pump 2. This combination was sent to the TE_1 input of the PIC, while pump 1 was connected to TE_0. The pumps are amplified by an Erbium-Doped Fiber Amplifier (EDFA)/Praseodymium-Doped Fiber Amplifier (PDFA), power-controlled by a Variable Optical Attenuator (VOA) and filtered by a Tunable Optical Band-Pass Filter (T-OBPF). The PIC was temperature controlled and reverse biased with 20 V. The output coupling angle was optimized for flat CE. Then, pump and signal polarization as well as the fiber-to-chip coupling were optimized for high CE. From the TE_0 output, the converted signal (idler) is either sent through a noise-loading stage or over a 100 km SSMF link.

Due to unavailability of a coherent O-band receiver, a KK-Rx was used. It adds a polarization aligned offset CW carrier to the signal and uses their beating on a Photo Diode (PD) to retrieve the complex baseband in the digital domain. Afterwards, it can be processed by conventional Digital Signal Processing (DSP) [27]. By adjusting the CW carrier PDFA, the carrier-to-signal power ratio was maintained at 13.5 dB. The CW carrier wavelength was placed to create a guard band of 1.8 GHz between the carrier and the received signal. The T-OBPF in the KK-Rx were adjusted to minimize the BER.

1.4.1.2 Experimental Results

For the following experiments, the input powers at the PIC were measured to be 11.3 dBm for the signal, 24.0 dBm for pump 1, and 19.3 dBm for pump 2. Starting from simulated optimum values, the pump wavelengths were iteratively adjusted to maximize both bandwidth and CE. Eventually, pump wavelengths of 1540 nm and 1300 nm were selected for pump 1 and

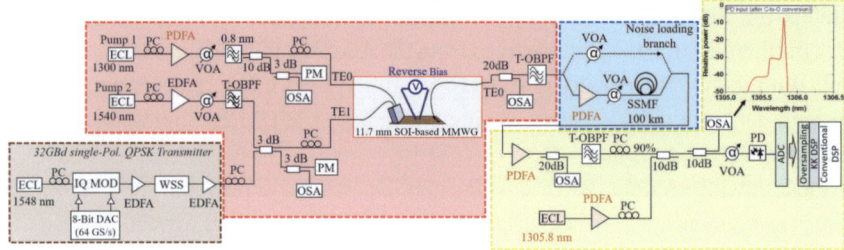

Fig. 1.35 Experimental setup for C → O conversion, including transmitter stage (brown box), AOWC stage (red box), noise-loading or optional transmission stage (blue box) and KK-Rx stage, including optical spectrum at PD input as inset (yellow box) [26]

pump 2, respectively, facilitating broadband operation with a 240 nm wavelength shift—a range rarely demonstrated in prior experimental work.

In the initial CW FWM characterization, the transmitter was replaced with a tunable External Cavity Laser (ECL). Figure 1.36a illustrates the CE results in these characterization measurements, as the ECL wavelength is swept from 1530 to 1565 nm, with all other conditions held constant. The peak CE measured was −41 dB, closely matching the simulated −39 dB, and a minimum 25 nm bandwidth was observed in the C-band. No efficiency roll-off was seen at longer wavelengths. When a pump within the C-band transmission window is acceptable, the entire 35 nm C-band bandwidth can be converted. Although experimental measurements were restricted to the C-band due to equipment limitations, PM simulations predict bandwidths approaching 100 nm (see Sect. 1.3.5). Nonetheless, in the current device, this is constrained by the ∼30 nm 1 dB bandwidth of the GCs. A chip design based on edge coupling could alleviate this restriction. The CE ripple is attributed to linear intermodal crosstalk.

C → O band conversion experiments were then conducted using data signals at 1542 nm, 1548 nm, and 1562 nm, with a Back-to-Back (B2B) reference measurement at 1548 nm. In the B2B scenario, the transmitter was directly connected to the noise-loading stage, and the ECL, PDFA, and T-OBPF in the KK-Rx were replaced by C-band equivalents. Results are shown in Fig. 1.36c, which indicates an OSNR penalty of less than 0.4 dB at the hard-decision forward error correction (HDFEC) threshold, compared to the B2B case. For consistency, OSNR values were corrected to a 12.5 GHz noise bandwidth.

In the 1548 nm case, the noise-loading setup was replaced by a 100 km SSMF transmission link with 35 dB IL. A VOA was used to adjust the launch powers of the converted signals. The resulting BER performance is shown in Fig. 1.36d, demonstrating that a launch power of only −11.5 dBm is sufficient to remain below the HDFEC threshold after 100 km.

According to Fig. 1.36b, pump 1 still exhibited considerable power at the AOWC output. Despite a T-OBPF suppressing this pump by approximately 30 dB, it remained dominant at the input of the following PDFA, thus constraining the allowable launch power of the

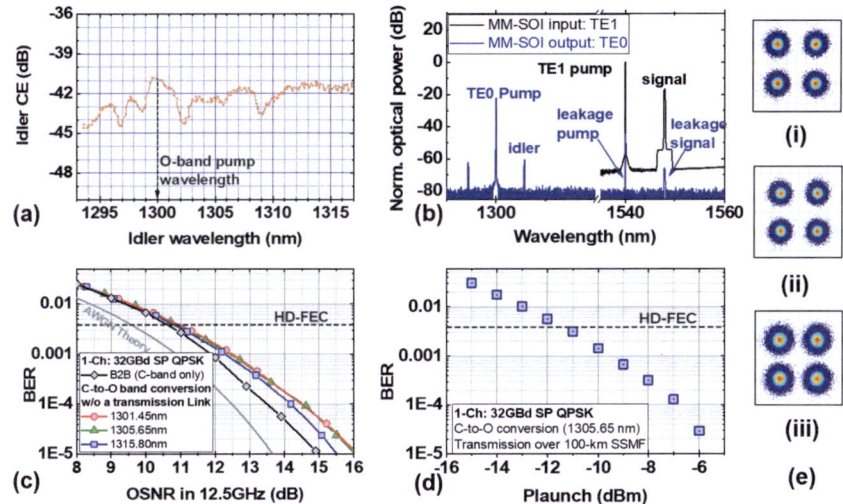

Fig. 1.36 a Experimental results of the AOWC, pumped by 24.0 dBm at 1540 nm and 19.3 dBm at 1300 nm: Measured CW CE for signals in the C-band. b Spectrum at the PIC input and output. c BER of 32 Gbd SP QPSK signals, received after noise-loading stage, and d after 100 km SSMF using different launch powers for a signal at 1548 nm. e Constellation diagrams after the transmitter (i), after the AOWC (ii) and after transmission over 100 km SSMF (iii) [26]

converted signal. Future implementations are expected to incorporate on-chip pump filtering to address this.

Figure 1.36e presents the received constellations after the transmitter, after the AOWC, and after 100 km of transmission, all recorded at the highest achievable OSNR. No significant degradation is observed after the AOWC stage, suggesting compatibility with higher order modulation formats.

These results collectively demonstrate that the integrated AOWC, capable of translating the full C band into the O band, operates effectively using intermodal FWM in a PIN diode assisted MMWG. A CE of −41 dB was achieved under moderate pump powers. Across all evaluated wavelengths, the AOWC introduced an OSNR penalty below 0.4 dB, and its performance was validated by successful transmission over 100 km of SSMF for a single channel. Leveraging a PIC platform also enabled the on-chip integration of MADM, offering a scalable, compact, and cost-efficient solution that would be impractical with conventional fiber-based architectures.

1.4.2 Multi-mode SOI Straight Waveguide for Ultra-Broadband C → C Optical Phase Conjugation

In this section, we present the use of a multi-mode SOI straight waveguide as a nonlinear medium to perform OPC. This device is specifically used as a mid-link element in a real optical fiber communication system to mitigate the signal data's nonlinearity and chromatic dispersion, improving the data quality after approximately 300 km of SSMF transmission.

The device used for OPC in this work is a straight NR waveguide based on SOI technology chosen for its high nonlinear coefficient, as discussed in Sect. 1.2.5. However, because silicon inherently experiences TPA, a PIN diode structure was incorporated into the design to reduce the free-carrier lifetime and mitigate the resulting attenuation by FCA. The general structure of the device is illustrated in Fig. 1.2. GCs are implemented at both ends to couple light into and out of the waveguide, while integrated Multiplexer (MUX) and Demultiplexer (DEMUX) structures are used to selectively excite and extract the TE_2 mode. This ensures mode-specific propagation along the waveguide, which is critical for maintaining the designed dispersion and enabling efficient OPC operation.

Figure 1.37 shows the schematic setup used for the mid-link OPC measurements in this work. It consists of a transmitter, three 52.8 km spans of SSMF forming the first half of the link, the OPC device, another three 52.8 km SSMF spans forming the second half of the link, and a digital coherent receiver. The transmitter was used to generate the desired data signals. As shown in Fig. 1.37, each fiber span is preceded by an EDFA and a VOA. The signal was first launched into the EDFA, which was operated at its optimal gain to minimize noise figure and maximize the SNR. The VOA was then used to control the launch power into the fiber spans. To monitor the signal throughout the setup, 20 dB optical splitters were placed after the EDFAs and VOAs. The main output branch, carrying 99% of the power, transmitted the signal into the fiber link, while the 1% branch was directed to an OSA for monitoring purposes.

1.4.3 Nonlinear Characterization

In a first step, the nonlinear characterization of the chip was performed using the setup shown in Fig. 1.38. A CW signal at 1543.87 nm and a pump at 1547.4 nm, both polarized to match the TE_2 mode of the waveguide, were combined using a WDM. In contrast to the theory in

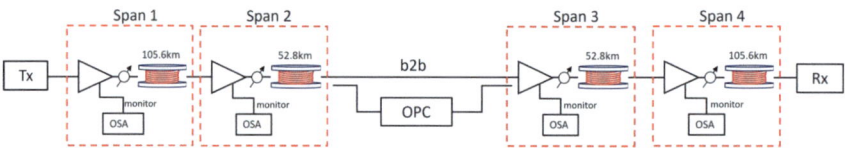

Fig. 1.37 Schematic design of an OPC setup over a 300 km SSMF link

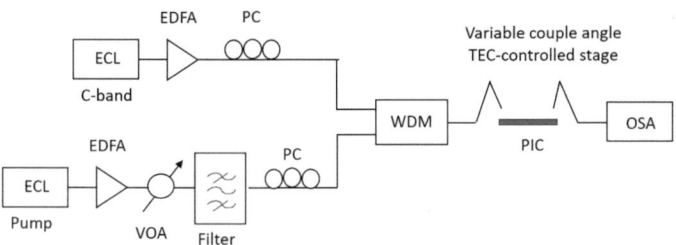

Fig. 1.38 Experimental OPC setup used for characterization of the OPC chip

Sect. 1.2, only one pump is implemented, i.e., $\omega_{P_1} = \omega_{P_2} = \omega_P$ in the OPC scenario. This results in an idler generation according to $\omega_I = \omega_P + \omega_P - \omega_S$. The maximum used pump power was at +25 dBm at the PIC inout to avoid damaging the GCs. The combined output from the WDM, containing both signal and pump, was coupled into the PIC via a grating coupler. To mitigate the effects of TPA and reduce free-carrier accumulation, a reverse bias voltage of 10 V was applied across the integrated PIN diode.

As shown in Fig. 1.40d, the idler generated through the FWM process is observed at 1550.95 nm in the OSA output. The input signal power was swept from −6 dBm to +18 dBm to evaluate the FWM efficiency and the idler generation while the pump wavelength was fixed at 1547.4 nm with a power of +25 dBm. As expected, the idler power increased linearly with signal power at lower input levels, indicating effective FWM interaction (see Fig. 1.40b). However, at higher signal powers, a saturation in the idler power was observed. This saturation can be attributed to several factors. Despite the reverse bias, residual TPA still generates free carriers, and at high optical intensities, the sweep-out of these carriers becomes less efficient, leading to increased FCA. Additionally, thermal effects and pump depletion may also contribute to this behavior by altering the PM conditions and reducing the available pump energy, respectively. These effects could limit the FWM CE [28].

The efficiency of signal conversion can be ultimately defined by the idler power at the waveguide output over the signal power at the waveguide input (see Eq. (1.25)). Figure 1.40c presents the idler power CE as a function of the input signal power. In this experiment, the signal wavelength was set to 1543.87 nm, while the pump was fixed at 1547.4 nm with a power of +25 dBm. The CE remains relatively constant about −39 dB for a broad range of signal powers. However, at higher input signal powers, a noticeable linear decrease in CE is observed. This decline at elevated signal powers is consistent with the saturation behavior discussed earlier. To summarize, the nonlinear characterization of the OPC device provided key insights into the behavior and limitations of the FWM process under varying optical power levels. It was found that, to prevent damage to the GCs, the pump power should be limited to a maximum of +25 dBm. Similarly, to avoid degradation in CE due to nonlinear distortion or saturation, the signal power should not exceed +16 dBm.

Under these optimized conditions, the device demonstrated a maximum CE of approximately −39 dB. These findings help define safe and effective operating conditions for the

OPC device and establish a solid foundation for system-level integration and performance optimization.

1.4.4 System Experiment Results

After evaluating the linear and nonlinear characterizations, we moved on to the system-level validation of the device. To quantify the OSNR penalty associated with the transmitter and receiver hardware for a 16QAM data signal, a noise-loading setup, realized with a VOA and subsequent EDFA at the receiver input was employed. The experimental setup is depicted in Fig. 1.39. The BER was then measured across a range of OSNR values. Figure 1.40e presents the results of BER performance measurements for a 1-channel and 3-channel 32 GBd, SP 16QAM signal in a B2B transmission setup without OPC, and for the scenario with OPC but still without the fiber link. The dashed line represents the theoretical BER curve derived from the AWGN model. Measurements were carried out for three cases: A single-channel transmission centered at 1543.87 nm (with and without OPC), and a WDM configuration using three channels at 1543.07 nm, 1543.87 nm, and 1544.67 nm, with 100 GHz channel spacing (without OPC). As observed from Fig. 1.40e at the HD-FEC threshold of BER = 3.8×10^{-3}, the single-channel case with OPC shows an OSNR penalty of approximately 0.5 dB compared to the measurements without OPC. The three-channel WDM case without OPC adds likewise a 0.5 dB OSNR penalty to the single-channel case, likely due to linear inter-channel crosstalk.

In the final experiment, the OPC stage was incorporated into the middle of a 300 km SSMF transmission link, as illustrated schematically in Fig. 1.37 and in more details in

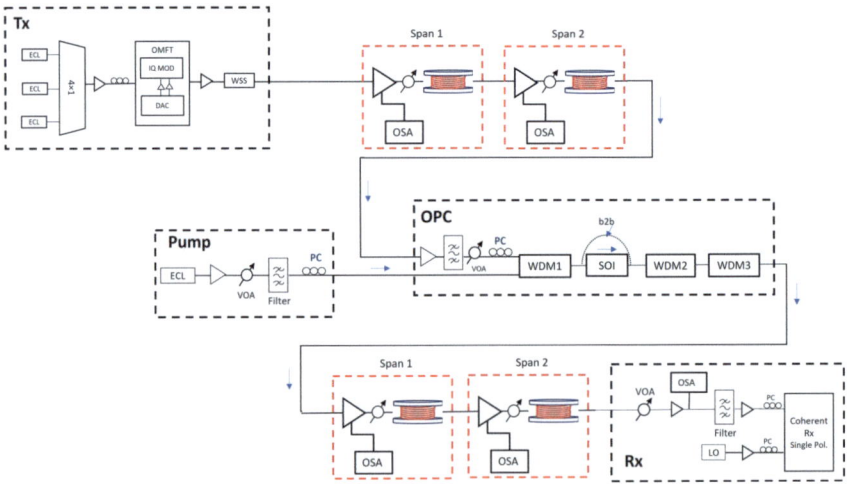

Fig. 1.39 Experimental setup used for OPC system measurements over a 300 km SSMF link

1 Ultra-Broadband Photonic Signal-Processor

Fig. 1.39. Figure 1.40f shows the BER performance as a function of launch power, ranging from −4 dBm to +8 dBm. This measurement was performed under two conditions: with and without the OPC stage. Without OPC, the CD of the link has to be compensated in the signal-processing unit of the coherent receiver. With OPC, CD is inherently compensated by the link itself. At launch powers below 0 dBm, the system operates in the linear regime and increasing the power improves BER. However, at higher launch powers, nonlinear effects become significant, leading to BER degradation. With the OPC stage, we observe an improvement in BER by one to two orders of magnitudes in the nonlinear regime. This clearly demonstrates the OPC's effectiveness in mitigating both nonlinear distortion and chromatic dispersion.

To conclude, the results from system-level measurements provide clear evidence that the multi-mode waveguide structure (enabling OPC) can mitigate the adverse effects of nonlinearities and dispersion in transmitted optical signals. However, one notable challenge identified during the experiment was the significant loss of 10 dB (see Fig. 1.40a) caused by the GC used for signal coupling into and out of the waveguide. This loss and the loss of the waveguide itself contribute to a considerable reduction in the OSNR, which limits the overall performance of the system. Despite this, the measurements serve as proof of concept for

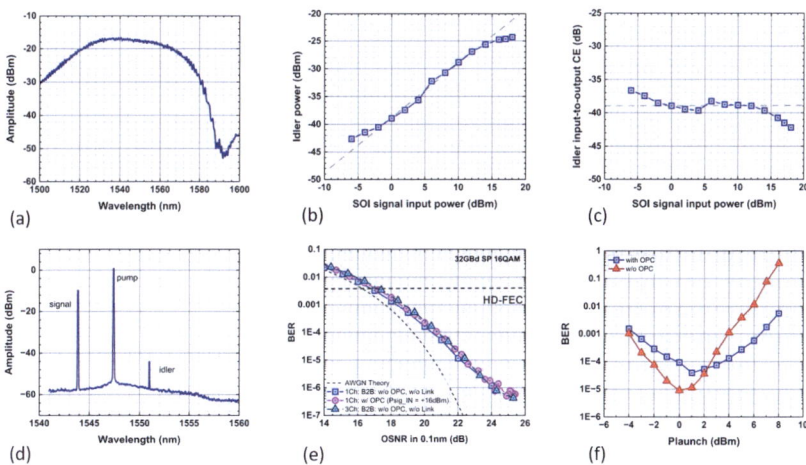

Fig. 1.40 **a** Calculated loss of a single grating coupler, based on B2B reference measurements. **b** Measured idler power as a function of input signal power, and **c** resulting efficiency of the FWM process. **d** Measured optical spectrum of the pump, signal, and generated idler wavelengths, demonstrating successful FWM in the OPC device. **e** BER performance versus OSNR for a 32 GBd SP 16QAM signal in a B2B configuration. Results are shown for both a single-channel case with and without OPC (1543.87 nm, pink circles and blue squares) and a three-channel WDM setup with 100 GHz spacing (blue triangles). The dashed line represents the theoretical BER curve based on the AWGN model. **f** BER versus launch power for a 300 km link with a 32 GBd SP 16QAM signal in a single-channel configuration. Results are shown for the case with and without the OPC stage

OPC using multimodal SOI waveguides, showing its potential for improving signal quality and enabling advanced optical processing techniques. The promising results indicate that the core design is solid, but further optimization is required to overcome the challenges posed by the coupling losses. Future work will focus on addressing these limitations by exploring the use of other high nonlinearity materials for the waveguide core, which could further enhance the performance of the system. Additionally, investigating edge coupling techniques holds potential for significantly improving the OSNR and minimizing losses, making it a promising avenue for future development. Overall, this work lays the foundation for continued research into optimizing multi-mode silicon waveguides for enhanced optical communication systems, opening new opportunities for ultra-broadband signal processing.

References

1. Hanik N, Kernetzky T, Jia Y, Höfler U, Freund R, Schubert C, Sackey I, Ronniger G, Zimmermann L (2022) Ultra-broadband optical wavelength-conversion using nonlinear multi-modal optical waveguides. In: 2022 13th international symposium on communication systems, networks and digital signal processing (CSNDSP). IEEE, pp 832–835
2. Winzer PJ, Neilson DT (2017) From scaling disparities to integrated parallelism: a decathlon for a decade. J Lightwave Technol 35(5):1099–1115
3. Maher R, Croussore K, Lauermann M, Going R, Xu X, Rahn J (2017) Constellation shaped 66 gbd dp-1024qam transceiver with 400 km transmission over standard smf. In: 2017 European conference on optical communication (ECOC). IEEE, pp 1–3
4. Fehenberger T, Böcherer G, Alvarado A, Hanik N (2015) Ldpc coded modulation with probabilistic shaping for optical fiber systems. In: Optical fiber communication conference. Optica Publishing Group, pp Th2A–23
5. Cho J, Chen X, Chandrasekhar S, Raybon G, Dar R, Schmalen L, Burrows E, Adamiecki A, Corteselli S, Pan Y et al (2017) Trans-atlantic field trial using probabilistically shaped 64-qam at high spectral efficiencies and single-carrier real-time 250-gb/s 16-qam. In: Optical fiber communication conference. Optica Publishing Group, pp Th5B–3
6. Kernetzky KT (2023) Numerical optimization of ultra-broadband wavelength conversion in nonlinear optical waveguides. PhD thesis, Technische Universität München
7. Kawano K, Kitoh T (2001) Introduction to optical waveguide analysis: Solving maxwell's equation and the schrödinger equation. Wiley, New York
8. Fallahkhair AB, Li KS, Murphy TE (2008) Vector finite difference modesolver for anisotropic dielectric waveguides. J Lightwave Technol 26(11):1423–1431
9. Agrawal GP (2012) Fiber-optic communication systems. Wiley, New York
10. Agrawal GP (2000) Nonlinear fiber optics. In: Nonlinear science at the dawn of the 21st century. Springer, pp 195–211
11. Rademacher G, Luís RS, Puttnam BJ, Furukawa H, Maruyama R, Aikawa K, Awaji Y, Wada N (2018) Investigation of intermodal four-wave mixing for nonlinear signal processing in few-mode fibers. IEEE Photonics Technol Lett 30(17):1527–1530
12. Kernetzky T, Jia Y, Hanik N (2020) Multi dimensional optimization of phase matching in multimode silicon nano-rib waveguides. In: 21th ITG-symposium photonic networks. VDE, pp 1–8
13. Essiambre R-J, Mestre MA, Ryf R, Gnauck AH, Tkach RW, Chraplyvy AR, Sun Y, Jiang X, Lingle R (2013) Experimental investigation of inter-modal four-wave mixing in few-mode fibers. IEEE Photonics Technol Lett 25(6):539–542

14. Osgood R Jr, Panoiu N, Dadap J, Liu X, Chen X, Hsieh I-W, Dulkeith E, Green W, Vlasov YA (2009) Engineering nonlinearities in nanoscale optical systems: physics and applications in dispersion-engineered silicon nanophotonic wires. Adv Opt Photonics 1(1):162–235
15. Höfler U, Kernetzky T, Hanik N (2022) Analysis of material susceptibility in silicon on insulator waveguides with combined simulation of four-wave mixing and linear mode coupling. Opt Quant Electron 54(12):837
16. Höfler U, Kernetzky T, Hanik N (2021) Modeling material susceptibility in silicon for four-wave mixing based nonlinear optics. In: 2021 International Conference on Numerical Simulation of Optoelectronic Devices (NUSOD). IEEE, pp 121–122
17. Gajda A, Zimmermann L, Bruns J, Tillack B, Petermann K (2011) Design rules for p-i-n diode carriers sweeping in nano-rib waveguides on SOI. Opt Express 19:9915–9922
18. Gajda A, Zimmermann L, Jazayerifar M, Winzer G, Tian H, Elschner R, Richter T, Schubert C, Tillack B, Petermann K (2012) Highly efficient CW parametric conversion at 1550 nm in SOI waveguides by reverse biased p-i-n junction. Opt Express 20:13100–13107
19. Tian H, Winzer G, Gajda A, Petermann K, Tillack B, Zimmermann L (2012) Fabrication of low-loss SOI nano-waveguides including BEOL processes for nonlinear applications. J Eur Opt Soc-Rapid Publ 7:12032
20. Zimmermann L, Thomson DJ, Goll B, Knoll D, Lischke S, Gardes FY, Hu Y, Reed GT, Zimmermann H, Porte H (2013) Monolithically integrated 10Gbit/sec Silicon modulator with driver in 0.25 μm SiGe:C BiCMOS. In: Proceedings of 39th European conference and exhibition on optical communication (ECOC), pp 1–3
21. Lischke S, Knoll D, Mai C, Zimmermann L (2019) Advanced photonic BiCMOS technology with high-performance Ge photo detectors. In: Mitrofanov O (ed) Optical sensing, imaging, and photon counting: from X-rays to THz 2019, vol 11088. International Society for Optics and Photonics, SPIE, p 110880M
22. Oton C, Lemonnier O, Fournier M, Kopp C (2016) Adiabatic bends in silicon multimode waveguides. In: 2016 IEEE 13th international conference on group IV photonics (GFP), pp 108–109
23. Ding Y, Xu J, Ros FD, Huang B, Ou H, Peucheret C (2013) On-chip two-mode division multiplexing using tapered directional coupler-based mode multiplexer and demultiplexer. Opt Express 21:10376–10382
24. Sanchez JE (2022) Devsim: a tcad semiconductor device simulator. J Open Source Softw 7(70):3898
25. Kernetzky T, Ronniger G, Höfler U, Zimmermann L, Hanik N (2021) Numerical optimization and cw measurements of soi waveguides for ultra-broadband c-to-o-band conversion. In: 2021 European conference on optical communication (ECOC). IEEE, pp 1–4
26. Ronniger G, Sackey I, Kernetzky T, Höfler U, Mai C, Schubert C, Hanik N, Zimmermann L, Freund R, Petermann K (2021) Efficient ultra-broadband c-to-o band converter based on multimode silicon-on-insulator waveguides. In: 2021 European conference on optical communication (ECOC). IEEE, pp 1–4
27. Sackey I, Elschner R, Schmidt-Langhorst C et al (2018) Practical trade-offs for kramers-kronig reception. In: 19th ITG-symposium photonic networks
28. Srivastava AK, Fejer MM, Litchinitser NM, Bruce AJ, DiGiovanni DJ (2000) Ultradense wdm transmission in l-band. IEEE Photonics Technol Lett 12:1570–1572

Open Access This chapter is licensed under the terms of the Creative Commons Attribution 4.0 International License (http://creativecommons.org/licenses/by/4.0/), which permits use, sharing, adaptation, distribution and reproduction in any medium or format, as long as you give appropriate credit to the original author(s) and the source, provide a link to the Creative Commons license and indicate if changes were made.

The images or other third party material in this chapter are included in the chapter's Creative Commons license, unless indicated otherwise in a credit line to the material. If material is not included in the chapter's Creative Commons license and your intended use is not permitted by statutory regulation or exceeds the permitted use, you will need to obtain permission directly from the copyright holder.

2 Optic-Electronic-Optic Interferometers for Ultrabroadband Arbitrary Digital Signal Processing

Sebastian Randel, Md. Salek Mahmud, Alexander Schindler, Patrick Runge and Martin Schell

Abstract

This chapter presents the development and experimental validation of an optic-electronic-optic (OEO) interferometer supporting flexible coherent add-drop operations. The system leverages coherent detection, real-time digital signal processing, and in-phase/quadrature (I/Q) electro-optic remodulation within one interferometer arm to enable programmable manipulation of the optical field. We demonstrate successful add-drop operation of a quadrature phase-shift keying (QPSK) signal and quantify system performance under various conditions. In addition to single-channel demonstrations, we present new experimental results using intradyne detection in a dense wavelength-division multiplexing (WDM) scenario, showing scalability to multi-carrier systems. Long-term operational stability is ensured by continuous phase tracking implemented in the digital signal processing (DSP) chain. These results highlight the potential of the OEO architecture for

---

S. Randel (✉) · Md. S. Mahmud
Institute of Photonics and Quantum Electronics, Karlsruhe, Germany
e-mail: sebastian.randel@kit.edu

P. Runge · M. Schell
Fraunhofer Heinrich-Hertz-Institute, Berlin, Germany

A. Schindler
Institute of Solid-State Physics, TU Berlin, Berlin, Germany

future coherent optical networks requiring dynamic channel control, integrated processing, and signal reconfigurability. Two generations of indium phosphide (InP)-based integrated OEO interferometer PICs are introduced to improve phase and polarization stability. As a preliminary step, a partially integrated OEO interferometer PIC is developed and assembled into a module. Finally, a fully integrated OEO interferometer PIC is presented, incorporating all optical components required for the OEO interferometer setup.

Acronyms

ADC	Analog-to-Digital Converter
ASE	Amplified Spontaneous Emission
AWGN	Additive White Gaussian Noise
BER	Bit-Error Ratio
DAC	Digital-to-Analog Converter
DPLL	Digital Phase-Locked Loop
DSP	Digital Signal Processing
ECL	External-Cavity Laser
EDFA	Erbium-Doped Fiber Amplifier
EO	Electro-Optic
EQ	Equalizer
ER	Extinction Ratio
ESA	Electrical Spectrum Analyzer
FIR	Finite Impulse Response
ICR	Integrated Coherent Receiver
I/Q	In-Phase/Quadrature
LO	Local Oscillator
LUT	Look-Up Table
MA	Moving Average
MIMO	Multiple-Input Multiple-Output
MZI	Mach-Zehnder Interferometer
MZM	Mach-Zehnder Modulator
NCO	Numerically Controlled Oscillator
OBP	Optical Bandpass
ODL	Optical Delay Line
OEO	Optic-Electronic-Optic
OQAM	Offset Quadrature Amplitude Modulation
OSA	Optical Spectrum Analyzer
OSNR	Optical Signal-to-Noise-Ratio
PC	Polarization Controller
PIC	Photonic Integrated Circuits

QPSK	Quadrature Phase-Shift Keying
SGDBR	Sampled Grating Distributed Bragg Reflector
SMF	Single-Mode Fiber
TIA	Trans-Impedance Amplifier
TWE	Traveling Wave Electrode
VOA	Variable Optical Attenuator
WDM	Wavelength-Division Multiplexing

2.1 Introduction

In recent years, advancements in coherent optical transceivers have been driven by the development of high-speed digital-to-analog converters (DACs), analog-to-digital converters (ADCs), and real-time digital signal processing (DSP) systems. These technologies have significantly improved bandwidth, efficiency, and cost-effectiveness, primarily benefiting coherent optical communications. However, their potential for hybrid optical-electronic processing is gaining increasing attention. By combining the high bandwidth of optics with the computational flexibility of DSP, new architectures can be designed for more versatile and adaptive signal processing.

A growing set of photonic-assisted methods is being explored to overcome the inherent bandwidth limitations of conventional electronics, enabling applications such as high-speed waveform synthesis, wideband optical sampling, and real-time signal conditioning [1–4]. In this context, we investigate the optic-electronic-optic (OEO) interferometer architecture, which enables dynamic and flexible manipulation of optical waveforms by converting them into the electrical domain, applying DSP-based modifications, and reconverting them into optics while preserving phase and polarization coherence. Conventional interferometers, based on transparent optical paths, are often limited to basic operations like phase shifts and time delays, enabling applications such as electro-optic modulators [5] and DPSK decoders [6]. In contrast, OEO architectures allow much more advanced processing and control of the optical waveform. This makes the OEO interferometric structure particularly well-suited for coherent add-drop multiplexing, where channels can be selectively extracted, modified, or reinserted. Unlike traditional wavelength-selective multiplexers that operate on fixed channel grids, OEO-based systems can handle baud-rate-spaced or even spectrally overlapping signals [7]. The method relies on coherent detection, digital channel separation, and electro-optic modulation for interferometric recombination. Beyond optical communications, OEO-interferometric superposition is also relevant for applications requiring phase-coherent feedback, such as coherent Ising machines [8]. By bridging the optical and electronic domains, these hybrid processing techniques pave the way for highly adaptive and reconfigurable photonic platforms, unlocking new possibilities in optical computing and signal processing.

Recent advances in interferometric add-drop multiplexing have explored different techniques, including all-optical processing via a gated fast Fourier transform (FFT) [9] and coherent detection followed by recombination through frequency conversion in a highly nonlinear fiber [10]. However, these two approaches differ fundamentally from the OEO interferometer, first proposed in [11, 12] and later studied through numerical simulations in [13]. The first experimental realization of an OEO interferometer was demonstrated in [14] utilizing analogue electrical processing to manipulate intensity-modulated signals. A piezoelectric fiber stretcher was employed to compensate for phase fluctuations, but its limited tuning range constrained long-term stability [15, 16]. This issue was addressed in [17] by incorporating a real-time DSP-based processing path capable of supporting complex modulation formats. Additionally, phase correction was implemented using an in-phase/quadrature (I/Q) modulator, enabling continuous phase tuning without range limitations.

In this work, we build upon our previous studies [17, 18] by presenting a generalized framework and a more streamlined implementation of the OEO interferometer. We provide a detailed overview of the necessary DSP steps and an in-depth analysis of the phase-control loop, demonstrating its ability to maintain long-term phase stability. Additionally, we describe the experimental implementation using an initial prototype integrated module and the design of the photonic-integrated circuits (PIC). The FPGA-based testbed enables real-time add-drop multiplexing functionality, and we present bit-error ratio (BER) measurements for a three-channel WDM signal with 2.9 GHz spacing and 2 GBd QPSK modulation. These results demonstrate the potential of OEO-based coherent optical multiplexers to bridge the photonic and electronic domains through dense integration with advanced CMOS technologies.

2.2 Optic-Electronic-Optic Interferometer

The fundamental components of an OEO Mach-Zehnder interferometer (MZI) are depicted in Fig. 2.1. In conventional MZIs, the signal carrier frequency remains unchanged in both arms. However, in an OEO MZI, in one arm, the signal is first downconverted to the electronic baseband, and after the desired processing it is converted back to the original carrier frequency. As shown in Fig. 2.1, the lower path of the OEO structure features coherent reception, DSP, and I/Q modulation. Insets show characteristic amplitude spectra at various stages of the signal path with the exemplary objective to replace an input optical channel at carrier frequency f_i with a different signal at the same frequency while keeping the neighboring channels unaffected.

Consider a WDM signal input to the OEO interferometer, with the optical field expressed as,

$$\underline{E}_{\text{in}}(t) = \sum_k \underline{a}_k(t) \exp\left(j[2\pi f_k t + \phi_k(t)]\right)$$

2 Optic-Electronic-Optic Interferometers for Ultrabroadband …

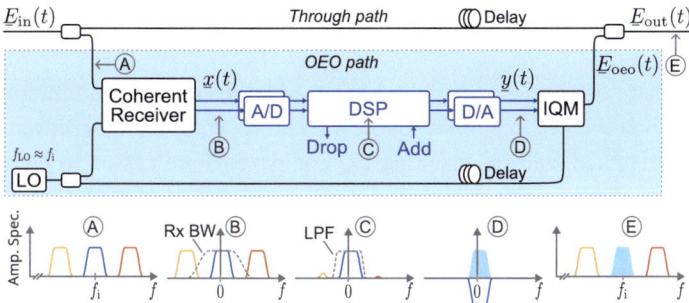

Fig. 2.1 Basic structure and building blocks of an optic-electronic-optic interferometer. Amplitude spectra at key stages: (A) input WDM signal, (B) baseband after reception, (C) after DSP filtering. Receiver and LPF bandwidths shown as gray dashed lines. (D) Regenerated electrical spectrum with new waveform (shaded) and inverted original spectrum (solid). (E) Final optical spectrum after add-drop operation

where, $\underline{a}_k(t)$ represents the kth channel's complex amplitude carrying information, f_k denotes the carrier frequency, and $\phi_k(t)$ accounts for the associated laser's phase noise. At the input of the interferometer, this signal is split into an optically transparent upper path and an optically opaque processing path. These are referred to as the *through* path and the *OEO* path respectively. In the OEO path, a free-running local oscillator (LO) laser operating at frequency $f_{LO} \approx f_i$ with phase noise $\phi_{LO}(t)$ is used in conjunction with a coherent receiver to detect the I/Q components of the input optical field,

$$\underline{E}_{\text{in}}(t) \exp\left(-j[2\pi f_{LO} t + \phi_{LO}(t)]\right) * h_{\text{crx}}(t),$$

where $h_{\text{crx}}(t)$ represents the impulse response of the bandwidth-limited coherent receiver. To suppress unwanted spectral components from neighboring channels, a lowpass filter can be used before the ADC. After lowpass filtering the resulting electrical signal can be written as,

$$\underline{x}(t) \propto \underline{a}_i(t) \exp\left(j[2\pi \Delta f t + \Delta\phi(t)]\right),$$

where $\Delta f = f_{\text{tx}} - f_{\text{LO}}$ denotes the frequency offset, and $\Delta\phi(t) = \phi_{\text{tx}}(t) - \phi_{\text{LO}}(t)$ is the phase offset. Next, the resulting signal is digitized using analog-to-digital converters (A/D) and processed through a series of DSP operations to demodulate the drop data and to modulate the new add data. Figure 2.2a illustrates a general framework for the required DSP operations for add-drop operation in an OEO interferometer. Initially, the frequency offset Δf is compensated coarsely. Then, a receive (Rx) filter with an impulse response $h_{\text{rx}}(t)$ is applied to suppress any residual image from the adjacent channels and the signal signal is downsampled to one sample per symbol. After phase offset $\Delta\phi$ is estimated and corrected, the recovered noisy symbol sequence is directed to a local drop output for demodulation.

Simultaneously, a sign-inverted copy is combined with a locally generated sequence of add symbols.

Next, the phase offset $\Delta\phi$ is reapplied followed by upsampling and filtering through a transmit (Tx) filter with an impulse response $h_{tx}(t)$. Subsequently, the recovered frequency offset Δf is reapplied, and the resulting I/Q components are passed to digital-to-analog converters (D/A). If phase preservation is not required at the interferometer output, both the phase recovery and phase modulation steps can be bypassed as indicated by the red dashed lines in Fig. 2.2a.

For the experimental setup presented in Sect. 2.3.1, we implemented a simplified DSP chain, as shown in Fig. 2.2b. Here, we directly subtract the digitized version of the signal $\underline{x}(t)$ from the frequency-shifted add signal, where the frequency offset is manually adjusted. The final output signal, after applying the DSP operations from the two DSP chains discussed above can be expressed as,

$$\underline{y}_1(t) = \left[\underline{a}_{add}(t)\, e^{j\Delta\phi(t)} * h_{tx}(t)\right] e^{j2\pi\Delta f t} - \underline{x}(t) * h_{trx}(t)$$

and

$$\underline{y}_2(t) = \left[\underline{a}_{add}(t) * h_{tx}(t)\right] e^{j2\pi\Delta f t} - \underline{x}(t),$$

respectively. Here, for $\underline{y}_1(t)$, the combined filter response, $h_{trx}(t) = h_{rx}(t) * h_{tx}(t)$ is essential for suppressing out-of-band noise and for subchannel selection in case of (orthogonal) frequency-division or wavelength-division multiplexing. Alternatively, the transmit filter defines the pulse shape of the add signal.

Following the DSP steps and D/A conversion, the signal $\underline{y}(t)$ (either $\underline{y}_1$ or $\underline{y}_2$) is modulated onto a copy of the LO laser, using an electro-optic I/Q modulator with a lowpass response of $h_{iqm}(t)$. This remodulated optical field propagating towards the output coupler of the interferometer is subject to random phase drifts denoted as ϕ_{oeo}, results in

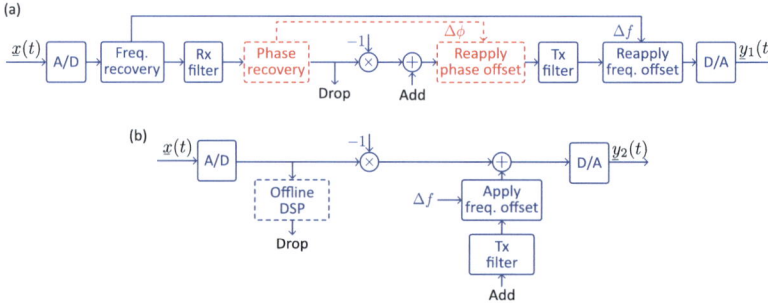

Fig. 2.2 Typical sequence of DSP steps in an optic-electronic-optic interferometer (**a**) and a simplified version implemented in Sect. 2.3.1b. Reprinted with permission from [18] © Optica Publishing Group

$$E_{\text{oeo}}(t) \propto \left[\underline{y}(t) * h_{\text{iqm}}(t)\right] e^{j[2\pi f_{\text{LO}}t + \phi_{\text{LO}}(t) + \phi_{\text{oeo}}]}.$$

At the output coupler, the signal from the OEO path is superimposed with the through path signal, which also experiences a random phase variation ϕ_{through}. Here, we assumed that the amplitudes, polarizations, and propagation delays of both interferometer arms are well-matched. Considering the two DSP implementations from Fig. 2.2 and neglecting the effects of $h_{\text{crx}}(t)$ and $h_{\text{iqm}}(t)$, the optical signals at the interferometer output can be expressed as,

$$E_{\text{out},1}(t) \propto \left(\underline{a}_{\text{in}}(t) e^{j\phi_{\text{through}}} - \left[\underline{a}_{\text{in}}(t) * h_{\text{trx}}(t)\right] e^{j\phi_{\text{oeo}}} + \left[\underline{a}_{\text{add}}(t) * h_{\text{tx}}(t)\right] e^{j\phi_{\text{oeo}}}\right)$$
$$\times e^{j[2\pi f_{\text{tx}}t + \phi_{\text{tx}}(t)]}$$

and

$$E_{\text{out},2}(t) \propto \left(\underline{a}_{\text{in}}(t) e^{j\phi_{\text{through}}} - \underline{a}_{\text{in}}(t) e^{j\phi_{\text{oeo}}} + \left[\underline{a}_{\text{add}}(t) * h_{\text{tx}}(t)\right] e^{j[\phi_{\text{oeo}} - \Delta\phi(t)]}\right)$$
$$\times e^{j[2\pi f_{\text{tx}}t + \phi_{\text{tx}}(t)]},$$

respectively. These expressions reveal that the compensation of the residual phase difference $\phi_{\text{through}} - \phi_{\text{oeo}}$ between the two paths is necessary to ensure effective signal cancellation.

The proposed OEO interferometer architecture can be extended to accommodate dual-polarization transmission systems. This would involve incorporating an adaptive multiple-input multiple-output (MIMO) equalizer, where the equalizer coefficients are inverted and applied to the add path. However, strong polarization-dependent loss could present challenges in practical implementations.

2.3 Experimental Implementations

This section details the experimental implementation of the OEO interferometer for coherent add-drop functionality. The experiments progress through increasingly advanced coherent detection setups. The initial setup employed coherent heterodyne detection with a single-channel input to establish the foundational DSP framework. Building on this, the add-drop operation is demonstrated using a preliminary integrated module. The final implementation then leverages intradyne detection to validate add-drop operation under dense WDM conditions.

2.3.1 Fiber-Based OEO Interferometer with Heterodyne Detection

The first experimental demonstration of the OEO interferometer is carried out for a single-channel input signal with coherent heterodyne detection. The goal is to validate the core DSP framework needed for real-time operation, including receiver-side signal recovery, transmitter-side signal generation for the added channel, and phase control for coherent interference. As illustrated in Fig. 2.3, the system comprises three main sections: a transmitter (Tx0) that generates the initial optical transmit signal, the central OEO interferometer responsible for the add-drop operation, and a coherent receiver (Rx0) used for signal analysis and performance evaluation.

A 1-GBd QPSK signal is synthesized within the FPGA fabric of a radio frequency system-on-chip (Xilinx RFSoC ZCU111). The process involves generating two delay-decorrelated pseudorandom binary sequences (PRBS) of length $2^{15} - 1$, mapping to QPSK symbols, and applying a pulse shaping filter using a root-raised-cosine frequency response with a roll-off factor of 0.4. Afterwards, the signal is converted to an analog waveform using two integrated 4-GHz D/A converters, sampling at a rate of 4 GSa/s. The resulting waveform occupies a baseband bandwidth of around 0.7 GHz. This design allows the use of a 1-GHz intermediate frequency for heterodyne detection in the OEO path, ensuring sufficient spectral separation between signal and image bands, while staying well within the Nyquist limit of the 4-GSa/s A/D converters. After amplification, the signal modulates an optical carrier at 193.489 THz provided by an external-cavity laser (ECL) with a linewidth of less than 100 kHz using a LiNbO$_3$ I/Q modulator (IQM).

At the interferometer input coupler, a 50/50 coupler splits the transmitted signal power into two distinct paths. In the lower path referred to as the OEO path, we perform balanced heterodyne detection utilizing a balanced photodetector module (BPD, Optilab BPR-20-M) and a free-running fiber laser (NKT) operating at 193.490 THz with 10 kHz linewidth serving as a local oscillator (LO1). The detected electrical signal is digitized using an A/D converter sampling at 4 GSa/s with an electrical 3-dB bandwidth of 4 GHz. From this real-valued digital signal, we form a complex-valued analytic signal by generating the imaginary component using the Hilbert transform ($\mathcal{H}$). The Hilbert transform is implemented by approximating it as a finite impulse response (FIR) filter in the time domain as outlined in [19, 20].

The analytic signal is next subtracted from a digitally frequency-shifted newly generated data signal, referred to as "Add data". The "Add data" is also a 1-GBd RRC pulse shaped

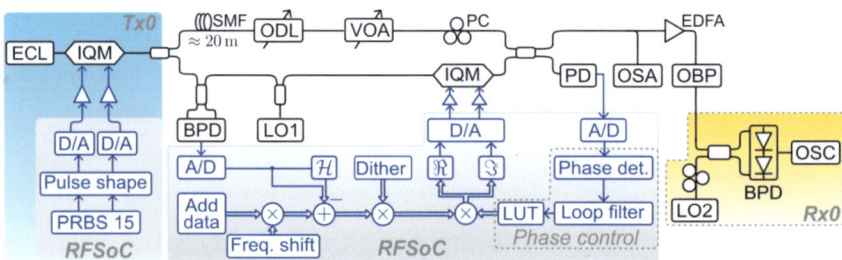

Fig. 2.3 Experimental setup of the coherent add-drop interferometer. The optical and electrical signal paths are visualized in black and blue, respectively. Within the DSP block (gray-shaded area), complex signals are indicated by a double arrow, while real-valued signals are shown with a single-line arrow. $\mathcal{H}$: Hilbert transform, $\Re$, $\Im$: real and imaginary part, respectively

QPSK waveform generated following the same process as before. In this implementation, we apply a fixed frequency shift of 1 GHz to nominally match the frequency difference of Tx0 and LO1. However, with the implementation of a real-time carrier recovery algorithm [21–24], the estimated carrier phase and frequency offset can be directly imposed onto the newly generated data signal.

Subsequently, a small-amplitude sinusoidal dither signal at a frequency of 15.6 MHz is applied, where the dither amplitude is quantized with 3-bit resolution compared to the 12-bit resolution of the received signal. The phase correction value from the phase control DSP block's look-up table (LUT) is then applied to this resulting signal to compensate for the random phase fluctuations between the two interferometer paths. Additional details on the phase control mechanism are discussed later in this section. Finally, the real ($\Re$) and imaginary ($\Im$) components of the signal are fed to two D/A converters, amplified, and used to modulate a portion of LO1 via a second $LiNbO_3$ IQM. The resulting remodulated optical signal is then combined with the signal from the upper path using another 50/50 coupler.

The superimposed signal from the upper output of this coupler is used for data evaluation and bit-error ratio (BER) measurements. First, the signal is amplified using an Erbium-doped fiber amplifier (EDFA), filtered using an optical bandpass filter (OBP) of bandwidth 10 GHz to reduce out-of-band amplified spontaneous emission (ASE) noise. An ASE noise source is then used to variably add ASE noise and vary the optical-signal-to-noise-ratio (OSNR). Another coherent receiver, denoted as Rx0, then performs balanced heterodyne detection of the signal. The receiver uses a free-running ECL (100 kHz linewidth) as a tunable local oscillator (LO2) and a 5-GHz bandwidth BPD (Thorlabs BDX3BA). The electrical signal is then sampled at 6.25 GSa/s using an oscilloscope (OSC, Tektronix DPO70808B) for offline evaluation. The recorded waveforms are first normalized and resampled to two samples per symbol. After that, symbol timing synchronization is achieved by applying a feedforward clock recovery [25]. Channel distortions are corrected using an adaptive equalizer based on a constant-modulus algorithm [26, 27]. This is followed by carrier recovery with the Viterbi & Viterbi algorithm before hard-decision demodulation [28].

For maximum suppression of the original signal through destructive interference, both interferometer paths must be well-matched in terms of group delay, polarization, amplitude, and phase. In our setup, we begin by coarsely adjusting the delay in the through path with roughly 20 m of single-mode fiber (SMF). A variable optical delay line (ODL) with a resolution of 0.3 ps is then used for fine adjustments. Amplitude and polarization are controlled using a variable optical attenuator (VOA) and a polarization controller (PC), respectively.

Despite these calibrations, the phase difference between the two paths fluctuates over time due to external disturbances, such as vibrations, temperature variations, and/or phase noise differences between the lasers in Tx0 and LO1. Consequently, active phase stabilization is required to maintain long-term stability. However, directly using interference-induced power variations as an error signal in a control loop leads to phase locking only within a limited range (0 to π or π to 2π) due to phase ambiguity, as power and phase do not have a one-to-one mapping. However, for proper operation of the OEO interferometer, it is essential to lock the phase at the interference extrema (maximum or minimum). To resolve this, we employ a dither-based approach [29, 30]. A small periodic phase modulation is introduced into one interferometer arm, and synchronous demodulation is used to extract the corresponding power variation derivative. The required phase correction is then applied via the second IQ modulator (IQM) in the OEO path, enabling continuous i.e., endless phase adjustment without range limitations [31, 32].

Figure 2.4 depicts the steps of computing the phase correction value in the PLL. First, a low-bandwidth photodetector (PD, DC –200 MHz) converts interference power variations into an electrical signal, serving as a feedback signal. Next, the signal is digitized by an A/D converter (bandwidth: 10 MHz to 1 GHz) then multiplied by the cosine component of the reference dither signal. The resulting output comprises three frequency components: one DC term, one component at the dither frequency ω_d, and/or another at $2\omega_d$. A moving average (MA) filter extracts only the DC component by eliminating the oscillatory terms, which serves as the phase error value. The phase error signal is downsampled by a factor of 1024 relative to the FPGA clock of 250 MHz and then fed to a proportional-integral (PI) controller (loop filter) [33]. The controller parameters are optimized using the Ziegler–Nichols method [34]. Finally, the loop filter output is mapped into a look-up table (LUT) that provides the complex-valued phase correction. The LUT functions as a numerically controlled oscillator around zero frequency, and due to its inherent modulo operation on the input index, continuous phase adjustment across the entire range of 0 to 2π is achieved.

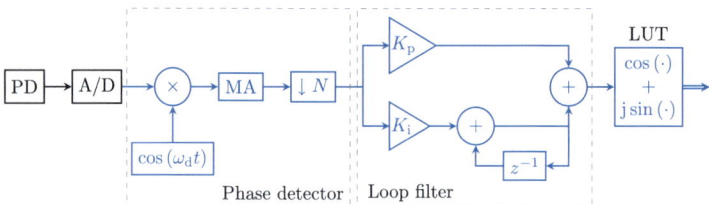

Fig. 2.4 Schematic of the phase control module consisting of a phase error detector, a loop filter, and a look-up table (LUT). The control loop includes a photodiode (PD) and an analog-to-digital converter (A/D) to measure interference power variation. An error signal around zero frequency is obtained by mixing with the reference dither and filtering the harmonics using a moving average (MA) filter. After downsampling ($\downarrow N$), the signal is fed into a proportional-integral (PI) loop filter to converge to the correct phase. The LUT converts the phase to a complex sinusoidal. Reprinted with permission from [18] © Optica Publishing Group

2.3.1.1 Results and Analysis

In the following, we evaluate the performance of the OEO interferometric system for channel suppression and reinsertion of a new data signal on the same channel, based on optical spectra and BER measurements.

Figure 2.5 shows optical spectra measured with a 20 MHz resolution bandwidth at the output coupler for five different scenarios. The solid blue trace represents the spectrum when only the through path is connected, while keeping the OEO path open. In contrast, the dashed red line corresponds to the OEO path's spectrum with the through path open. At the signal carrier frequency of 193.489 THz, the optical power levels are adjusted to achieve similar spectral intensity in both paths. It is visible that the OEO path is able to regenerate the original signal without any noticeable distortion. However, at 193.49 THz, the residual carrier from LO1 is visible alongside minor spectral broadening of the signal spectra caused by modulator nonlinearity. The digital Hilbert transform effectively suppresses the upper sideband near 193.491 THz by approximately 25 dB. When both interferometer paths are simultaneously connected and the controller is tuned for constructive interference (solid gray), the signal power increases by 6 dB, verifying phase coherence. Conversely, setting the controller to destructive interference yields the solid cyan spectrum, where the signal is suppressed by approximately 17 dB relative to the through-path-only scenario. Finally, when a new data signal is introduced while maintaining destructive interference, the resulting spectrum (dotted orange) exhibits a slight frequency shift, attributed to manual tuning of LO1 and inherent laser drift.

Figure 2.6a presents the temporal evolution of the monitor PD output voltage resulting from interference-induced power variation. The yellow line corresponds to the unstabilized

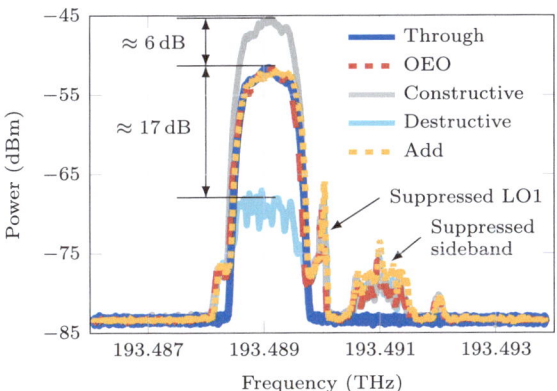

Fig. 2.5 Measured optical spectra (20 MHz resolution bandwidth) recorded at the interferometer output coupler for the upper (through) path and lower (OEO) path without any interference, for constructive and destructive interference, and for the case where the former signal is dropped and a new signal is added (add). Reprinted with permission from [18] © Optica Publishing Group

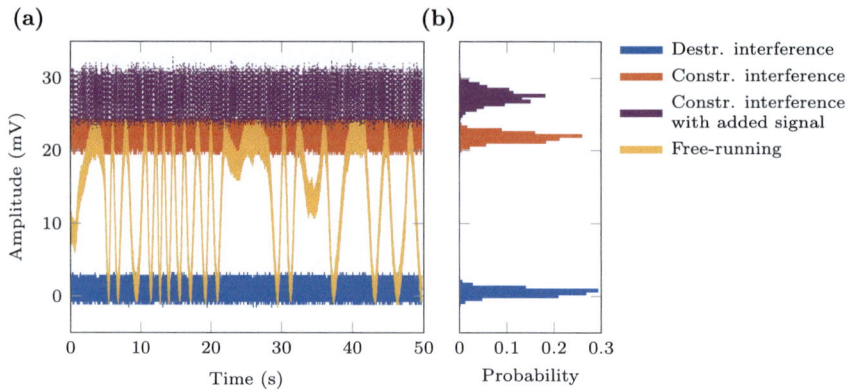

Fig. 2.6 Measured photodiode voltage over time (**a**) and corresponding histograms of the measured voltage signal (**b**). Reprinted with permission from [18] © Optica Publishing Group

interference condition, where the output voltage randomly oscillates between the upper and lower extremes. The blue and red curves indicate the actively stabilized PD output under destructive and constructive interference, respectively. When the new add-signal is introduced into the OEO path while maintaining constructive interference, the purple trace emerges, which produces the corresponding add spectrum depicted in Fig. 2.5. Since the two interferometer outputs have 180° phase offset, a destructive interference in the upper output corresponds to constructive interference in the lower output. The related histograms for the various voltage levels are displayed in Fig. 2.6b.

The BER performance of the add-drop process is examined under varying OSNR conditions. Amplified spontaneous emission (ASE) noise is injected after the OEO interferometer, and the OSNR is determined within 0.1 nm reference bandwidth. BER values are obtained for four distinct cases and compared with a theoretical benchmark (solid black), derived analytically for a 1 GBd QPSK signal over an additive white Gaussian noise (AWGN) channel, as plotted in Fig. 2.7. Due to ASE noise folding from the image band in heterodyne reception, an additional 3 dB penalty is included, in accordance with prior studies [35–38]. The BER measured using only the through path, shown by solid purple line, establishes a baseline for optical back-to-back characterization. BER measurements for the OEO path alone are first obtained without dither (dashed orange) and then with dither enabled (solid cyan), showing a sensitivity penalty of approximately 0.5 dB at a BER of 10^{-3}.

This penalty could potentially be mitigated by refining the bit resolutions of the DSP signals and/or reducing the dither frequency (see Fig. 2.3). When the new add-signal is enabled (dotted magenta) in place of the dropped channel, an additional 0.1 dB penalty is observed due to linear crosstalk from the residual drop signal. As a result, the total OSNR penalty introduced by the add-drop process, considering both the dither tone and crosstalk effects, is around 0.6 dB for QPSK modulation. The corresponding BER performance is

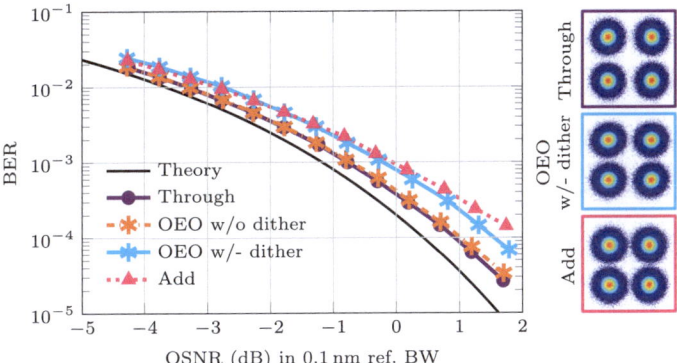

Fig. 2.7 BER measurements with varying OSNR for the transmitted QPSK signal before and after interferometric add-drop operation. The insets show the corresponding constellations (10^5 symbols plotted) for the cases of through, OEO with dither, and add. Reprinted with permission from [18] © Optica Publishing Group

supported by the constellation diagrams captured at 1.7 dB OSNR, shown in the inset, where the top-to-bottom order represents the through path, OEO with dither, and add-drop scenarios. These results confirm the viability of replacing an existing channel with a new signal while incurring minimal penalty, reinforcing its potential for future high-capacity optical networks.

2.3.2 Integrated OEO Interferometer Module with Heterodyne Detection

Implementing the interferometer system using a photonic-integrated circuit (PIC) will improve both polarization and phase stability. As an initial demonstration, passive waveguides, three 1×2 MMI couplers, and a 90° optical hybrid-based coherent receiver are monolithically integrated on an InP platform. The final PIC is co-packaged with a commercially available trans-impedance amplifier (TIA) and edge-coupled to fiber connectors.

The experimental setup, which includes the photonic module along with external delay and remodulation paths, is shown in Fig. 2.8. The components of the integrated module are highlighted within a red-shaded area. The transmit signal generation follows the same approach as described in the previous section. A Xilinx RFSoC (ZCU111) with integrated 4 GSa/s D/A converters produces a 1 GBd QPSK waveform, shaped by a root-raised cosine filter with a roll-off $\beta = 0.4$. This waveform then modulates an external cavity laser (ECL) using an IQ modulator (IQM).

The amplified optical signal enters the integrated module, splitting into a through path exiting the PIC and an OEO path to a coherent receiver. A free-running laser (LO1) (10 kHz line-width) is similarly split within the PIC. In our experiment with balanced heterodyne

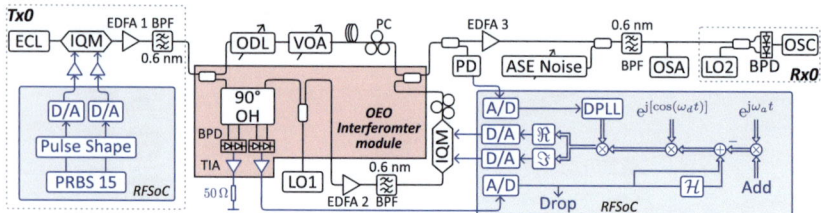

Fig. 2.8 Experimental setup using an integrated optic-electronic-optic interferometer module (ODL: Optical delay line, VOA: Variable optical attenuator, PC: Polarization controller, OBP: Optical bandpass filter). Reprinted with permission from [39] © Optica Publishing Group

detection, one BPD/TIA output is used while the other is terminated. The resulting electrical signal is digitized by a 4 GSa/s A/D converter on the RFSoC.

Building on the previously described DSP processing, we use the Hilbert Transform ($\mathcal{H}$) to generate an analytic signal and then subtract from it a digitally frequency-shifted 1 GBd QPSK signal, labeled "Add," obtained by multiplying with a complex exponential $\exp[j\omega_a t]$. Similar to the previous implementation, the frequency offset is ω_a is roughly 1 GHz. The next step involves multiplying the difference signal by a phase dithering term $\exp[j \cos(\omega_d t)]$ and the output of the digital phase-locked loop (DPLL). The real and imaginary components of the resulting signal then modulate a portion of LO1 using a second IQM. Both the remodulated optical signal and the through path signal are then routed back into the PIC, where they are combined and coupled out. The interferometer output is divided into two distinct paths: one is used for coherent reception (Rx0) and bit-error-rate (BER) analysis through offline DSP, while the other monitors optical power fluctuations induced by thermal and mechanical phase drifts with a low-bandwidth photodetector (PD) (DC–200 MHz). The phase error is extracted by the DPLL through a lock-in technique, as presented in the prior section, and phase correction values are calculated using a PI controller. In the external delay path, the group delay between the two interferometer arms is synchronized using patch cords and a variable optical delay line (ODL). A VOA and PC are then employed to adjust both the interference amplitude and polarization.

2.3.2.1 Results and Analysis

Following the methodology outlined earlier, we evaluate the performance of the interferometric add-drop operation using optical spectral analysis and BER measurements.

Figure 2.9 presents the optical spectra obtained under five distinct conditions, with a portion of the PD input tapped for measurement. The solid blue and dashed red traces represent the spectra of the individual interferometer through and OEO paths, respectively. The solid gray and solid cyan traces represent constructive and destructive interference, leading to a 6 dB power enhancement and an extinction ratio of 15 dB, respectively, compared to the individual paths. The dashed yellow trace represents the spectrum following the add process. Similar to the measurements in the earlier section, all spectra, except for the interferometer through path, exhibit suppression of LO1 near 193.49 THz and effective attenuation of the upper sideband near 193.491 THz, an effect of the digital Hilbert transform.

Figure 2.10 illustrates the measured BER as a function of OSNR for this integrated module-based setup. While the overall trends align with the previous full fiber-based setup measurements, slight increase in BER penalties are observed. The theoretical curve obtained analytically for an AWGN channel is shown in solid black. The through path (circle marker) again serves as the baseline measurement. The OEO path without dithering (dash-dotted orange line) exhibits a 0.2 dB penalty, while enabling the dither signal (solid cyan line) introduces an additional 0.6 dB penalty at a BER of 10^{-3}. Both penalties are slightly higher than those observed in the fiber-based measurement. For the add operation (triangle marker), the overall penalty amounts to 1.3 dB, which correlates with the 2 dB lower extinction ratio compared to the full fiber-based implementation. The increased penalty may result from length mismatch and polarization misalignment introduced during fiber-to-chip coupling, particularly due to the off-chip single-mode fiber delay in the through path.

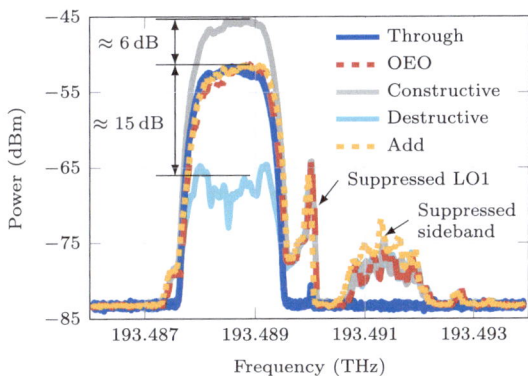

Fig. 2.9 Measured optical spectra (20 MHz resolution bandwidth) for individual through path and OEO path, for constructive and destructive interference and for the add-drop operation (Add). Adapted with permission from [39] © Optica Publishing Group

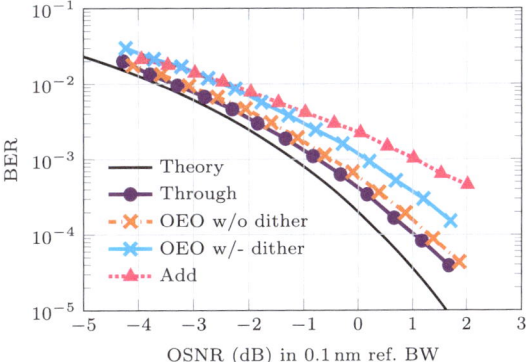

Fig. 2.10 BER performance for varying OSNR with the integrated interferometer module for single carrier heterodyne reception at the OEO path. Adapted with permission from [39] © Optica Publishing Group

In summary, we have demonstrated a functional OEO interferometer module based on an InP PIC incorporating passive waveguides and coherent receivers. The module achieves a coherent add-drop operation of 1-GBd QPSK signals with only a slight penalty increase compared to the fiber-based setup. This small additional penalty, caused by fiber-to-chip coupling issues, can be minimized through methods such as lensed fiber integration, which enhances mode field overlap with the chip's waveguide. With further refinement, such integrated solutions offer a promising path toward compact, scalable, and fully integrated interferometric optical add-drop multiplexing systems.

2.3.3 Fiber-Based OEO Interferometer with Intradyne Detection for WDM Signals

Building on the single-channel heterodyne detection based experiment, the setup is extended to support a multi-carrier WDM signal using intradyne detection. The experimental configuration is illustrated in Fig. 2.11. The transmitter (Tx0) now features an electro-optic frequency comb (EO-COMB) implemented using a Mach-Zehnder modulator (MZM), as detailed in the inset. The MZM is driven in push-pull configuration by a 2.9 GHz sinusoidal RF tone from a synthesizer. The drive voltage is tuned to produce three spectral lines, the carrier (0th order) and the first-order sidebands (± 1st). The MZM bias is further adjusted to equalize the power of these three spectral lines.

The comb lines are subsequently modulated by a 2-GBd QPSK waveform that has been pulse-shaped with an RRC frequency response having a roll-off factor of 0.1. Similar to the previous setup, the transmit signal is split into an upper through path and a lower OEO path. In the OEO path, coherent intradyne detection is performed using an integrated coherent receiver (ICR, Fujitsu FIM24902) and a narrow-linewidth (10 kHz) fiber laser (LO1) as the local oscillator. The resulting I/Q signals are digitized using two A/D converters and

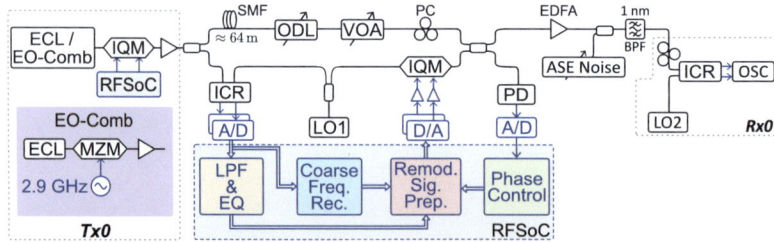

Fig. 2.11 Experimental setup of the coherent add-drop interferometer based on intradyne reception for WDM operation. Optical and electrical signal paths are represented by black and blue lines, respectively. Within the gray-shaded DSP block, complex-valued signals are denoted by double arrows, and real-valued signals by single-line arrows. ECL: External cavity laser, ODL: optical delay line PC: polarization controller, ICR: integrated coherent receiver

2 Optic-Electronic-Optic Interferometers for Ultrabroadband …

processed through a series of DSP stages. These operations are grouped into functional blocks, including a lowpass filter (LPF), an equalizer (EQ), a coarse frequency recovery block (Coarse Freq. Rec.), and a remodulation signal preparation stage (Remod. Sig. Prep.), which generates the waveform used to remodulate a portion of LO1 via an IQ-modulator. The internal structure and the signal flow of these DSP modules are discussed in detail later in this section.

A fiber delay of 64 m is added to the through path to coarsely align the group delay with the OEO path. As in the previous setup, one output of the interferometer is fed to a PD for phase control operation, while the other output is routed to a receiver (Rx0), employing a separate ECL (100 kHz linewidth) as LO (LO2) and another ICR, for performance evaluation after ASE noise loading. The received signal is then sampled using an oscilloscope (OSC) at 4 GSa/s and evaluated offline as described in Sect. 2.3.1.

The DSP architecture implemented in the OEO path is illustrated in Fig. 2.12, with the corresponding filter responses shown in Fig. 2.13. The received I/Q signal is processed along two parallel branches, each addressing different signal conditioning tasks. In the first branch, a coarse frequency offset compensation algorithm is employed. A 15-tap band-edge filter, whose frequency response is shown in Fig. 2.13a, is used to estimate the frequency offset between Tx0 and LO1 [40]. For efficient implementation, the filter response is decomposed into even and odd components, taking advantage of the properties that the imaginary part of the even filter and the real part of the odd filter are zero. The even filter output is delayed and multiplied with the conjugate of the odd filter output. The real part of the resulting product yields the frequency error estimate. This error signal is smoothed via moving average (MA), downsampled, and passed through a PI loop filter. The resulting control signal drives a numerically controlled oscillator (NCO), which synthesizes a complex exponential representing the estimated frequency offset. The correction signal is multiplied with the received I/Q waveform to complete the feedback path, while its conjugate is applied to the add-data signal to ensure frequency alignment.

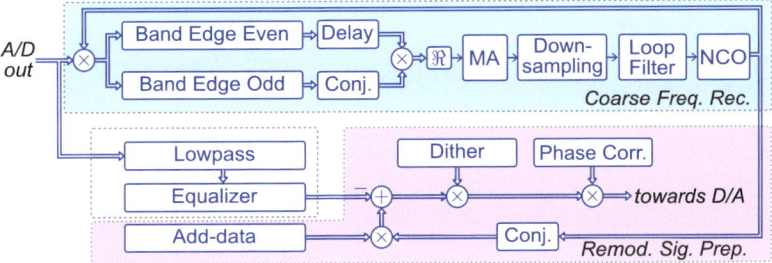

Fig. 2.12 Block diagram of the DSP chain in the OEO path supporting add-drop operation for WDM signals. Functional blocks for coarse frequency offset estimation, received signal correction, and remodulation signal synthesis are highlighted with blue, yellow, and red shading, respectively

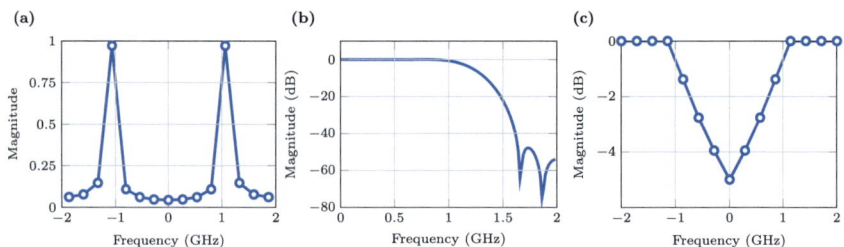

Fig. 2.13 Frequency responses of FIR filters used in the DSP chain: **a** 15-tap band-edge filter for frequency offset estimation; **b** lowpass filter designed using a Kaiser window; **c** 15-tap inverted-triangle filter response used for spectral shaping

In the second branch, the received signal is filtered using a Kaiser-windowed lowpass filter (Fig. 2.13b) to suppress out-of-band spectral components. This is followed by an equalizer, designed to mitigate spectral distortion introduced by the ICR and the second transmitter. The equalizer employs a 15-tap inverse-triangle spectral response, shown in Fig. 2.13c. The equalized signal is then subtracted from the frequency-shifted add-data signal. Following this, a phase-dither signal and a phase correction value from the phase control module (see Fig. 2.4) are applied. Finally, the real and imaginary components are routed to D/As to drive an IQ modulator for signal regeneration.

2.3.3.1 Results and Analysis

To assess the performance of the intradyne-based OEO interferometer, we first operate with a single carrier signal. This enables the evaluation of the ICR performance and allows careful optimization of the OEO path's DSP operations, particularly the equalizer, phase-locked loop, and the frequency-locked loop.

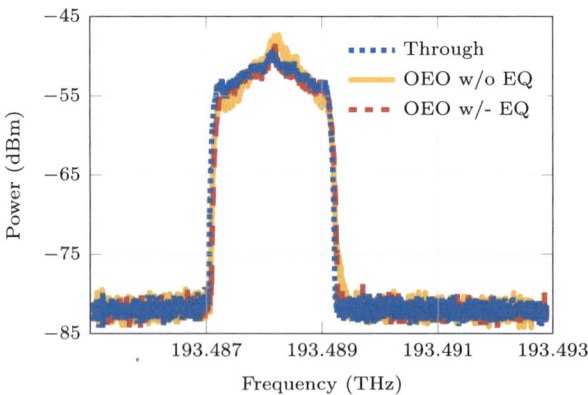

Fig. 2.14 Comparison of optical spectra for the through path (blue), the OEO path without equalization (yellow), and with the equalizer (red), measured with a resolution bandwidth of 20 MHz

Figure 2.14 illustrates optical spectra at the interferometer output coupler for the individual through and OEO paths, with the equalizer (EQ) either active or bypassed. The dotted blue trace corresponds to the through path and reflects only the spectral response of Tx0. A gradual linear reduction in spectral power from the center peak toward the band edges produces a characteristic triangular spectral envelope. This profile is likely caused by the non-flat low-frequency response of the DAC and RF amplifier. The solid yellow trace shows the OEO path spectrum without equalization, which exhibits a noticeably steeper decline in spectral power across the signal bandwidth. This spectrum reflects the combined frequency response of Tx0, the ICR, and the second transmitter in the OEO path. The enhanced roll-off is attributed to the cascaded effect of the two transmitters and the contributions from the elevated baseband noise introduced by the ICR's TIA. When the equalizer is enabled (dashed red trace), the OEO spectrum shows effective compensation for these distortions, resulting in a spectral shape comparable to the through path.

Figure 2.15 illustrates optical spectra, measured with the complete OEO DSP chain enabled in the intradyne configuration. The solid blue curve corresponds to the through path alone, whereas the dashed red line corresponds to the OEO path. As in the heterodyne case, the signal powers of both paths are carefully adjusted to yield comparable spectral intensities within the signal band. When both interferometer arms are connected and the phase controller is set for constructive interference (solid gray trace), a power enhancement of roughly 6 dB is observed, indicating phase coherence. In contrast, the solid cyan trace shows the spectrum under destructive interference, where the signal is suppressed by about 18 dB relative to the through-path-only case. These relative power levels are estimated by averaging power within a 2 GHz spectral extent centered around the peak. Residual leakage of LO1 is also evident at the center of the signal band. Lastly, introducing the add-data waveform, while the controller is set to destructive interference, produces the dotted yellow spectrum. The add-data signal is digitally scaled to match the power levels of the individual paths.

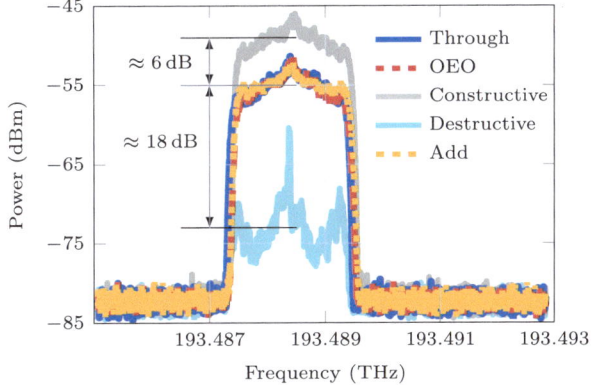

Fig. 2.15 Optical spectra (20 MHz resolution) of a 2 GBd QPSK signal recorded at the interferometer output for individual through/OEO paths, constructive/destructive interference, and add operation

The BER performance of the single carrier intradyne detection configuration is presented in Fig. 2.16. As in the heterodyne case, BER values are shown for four distinct cases, and OSNR is determined within 0.1 nm reference bandwidth. The analytic curve for a 2 GBd QPSK signal in an AWGN channel is plotted in solid black. The solid purple line with circle markers shows the BER for the through path alone, serving as an optical back-to-back reference. The OEO path without dither (dashed orange) and with dither (solid cyan) both exhibit a small 0.3 dB OSNR penalty at a BER of 10^{-3} due to the optical–electrical–optical conversion process. Notably, the addition of the dither signal introduces no further penalty. When the add-data signal is enabled (dotted magenta) in place of the dropped channel, a slight additional penalty of 0.1 dB is observed, attributed to linear crosstalk from the residual drop signal. Overall, the total OSNR penalty from the complete add–drop process including both OEO conversion and crosstalk effects is approximately 0.4 dB. The inset shows constellation diagrams measured at 2.75 dB OSNR, corresponding (top to bottom) to the through path, OEO with dither, and add-data cases.

Figure 2.17 presents the measured optical spectra for a three-channel WDM signal with 2.9 GHz channel spacing. The blue trace corresponds to the spectrum obtained from the through path alone, while the dashed red line shows the spectrum from the OEO path alone with the center channel reinserted via OEO remodulation. When both paths are combined with the controller tuned for constructive interference, the resulting spectrum (solid gray) shows a 6 dB increase in the peak power of the center channel, consistent with earlier single-channel results. Destructive interference is illustrated by the solid cyan curve, exhibiting approximately 17 dB suppression of the center channel relative to the individual paths. This suppression is quantified as the mean power reduction within a 2 GHz bandwidth centered around peak spectral power of the center channel. The outcome of the add operation applied to the center channel is shown by the dotted yellow trace. For comparison, Figs. 2.18

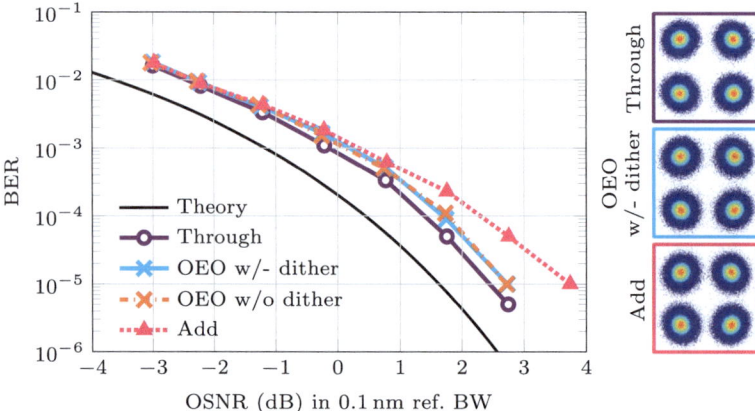

Fig. 2.16 BER versus OSNR for intradyne-detected single-carrier signal with corresponding constellations at 2.7 dB OSNR

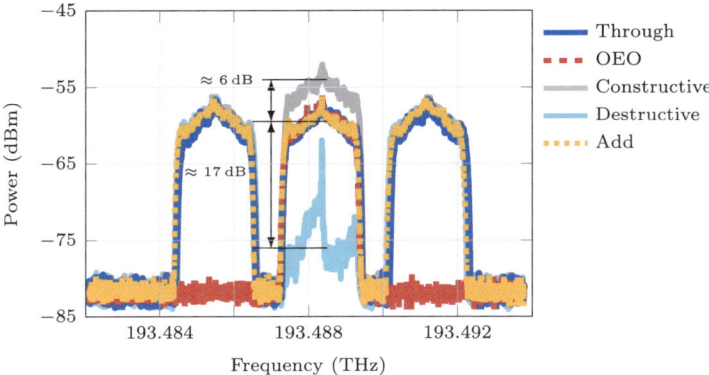

Fig. 2.17 Optical spectra at the interferometer output for the three channel WDM signal with add-drop of the center channel. The resolution bandwidth is 20 MHz

and 2.19 display the corresponding spectra when the add-drop functionality is applied to the left and right WDM channels, respectively. For the right channel, the destructive interference condition (Fig. 2.19) exhibits 1 dB lower extinction compared to the left or center channels. This reduction is attributed to increased residual leakage from LO1, likely stemming from bias drift in the IQ modulator of the OEO path.

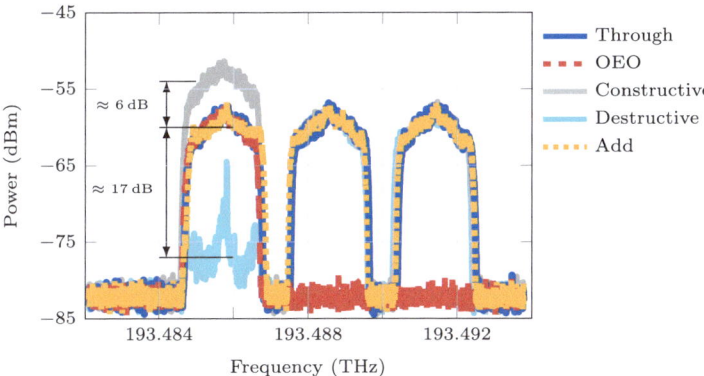

Fig. 2.18 Optical spectra at the interferometer output for the three channel WDM signal with add-drop of the left channel. The resolution bandwidth is 20 MHz

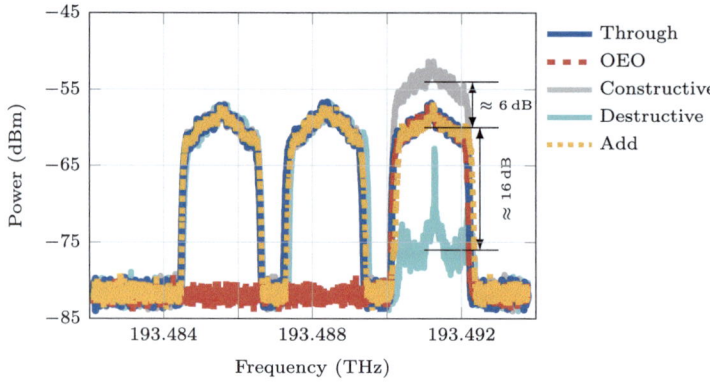

Fig. 2.19 Optical spectra at the interferometer output for the three channel WDM signal with add-drop of the right channel. The resolution bandwidth is 20 MHz

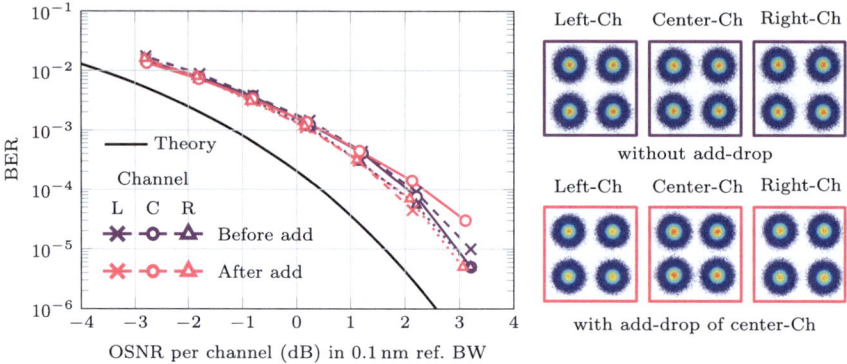

Fig. 2.20 BER measurements for three WDM channels before and after add-drop of the central channel (left). Corresponding constellations at 3 dB OSNR (right)

BER measurements for all three WDM channels were carried out before and after performing an add-drop operation on the center channel. The EO-Comb generator was adjusted to ensure equal peak power across all channels, and the optical spectrum analyzer (OSA) resolution bandwidth was set to 0.1 nm, wide enough to encompass the full three-channel WDM signal. The results are summarized in Fig. 2.20 alongside the analytic curve for 2 GBd QPSK signal. The measurement labeled *before add* (purple curve) corresponds to the case where only the through path was active. The *after add* operation case (magenta curve) reflects the scenario where a new data signal is inserted at the center wavelength using the OEO interferometer. At BER values above 10^{-3}, all three channels exhibit similar BER performance before and after the *add* operation. A minor penalty of approximately 0.4 dB is observed at lower BER values (e.g., at 10^{-4}) for the central channel after the add operation, attributed primarily to residual crosstalk from incomplete suppression during the drop stage.

This BER performance is further reflected in the constellation diagrams shown on the right of Fig. 2.20. These constellations, captured at 3 dB OSNR, illustrate the signal quality for the three WDM channels without (top) and with (bottom) the add-drop operation.

This performance trend is consistently observed when the add-drop operation is applied to the left and right WDM channels, as shown in Figs. 2.21 and 2.22, respectively. In particular, a slightly elevated BER is seen for the right channel after add-drop at OSNR levels above 2 dB, which correlates with the reduced extinction (by approximately 1 dB) during the drop stage, as illustrated in Fig. 2.19. These results validate the scalability of the OEO-based add-drop architecture to multi-channel systems, with consistent performance across all WDM channels and manageable crosstalk penalties.

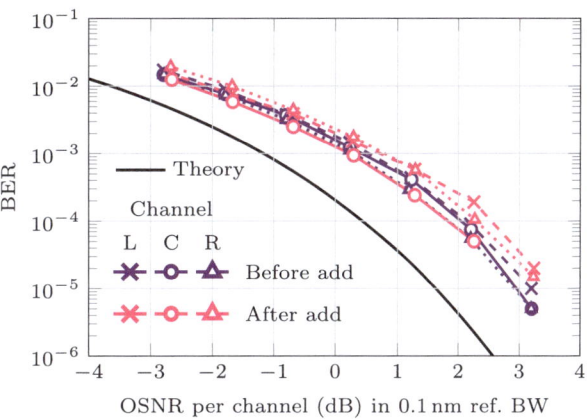

Fig. 2.21 BER measurements for three WDM channels before and after add-drop of the left channel

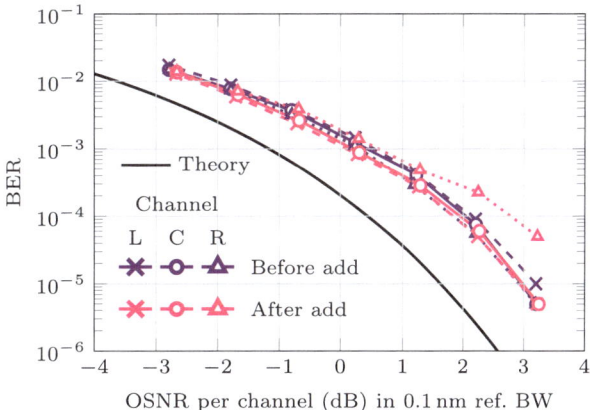

Fig. 2.22 BER measurements for three WDM channels before and after add-drop of the right channel

2.4 Monolithically Integrated OEO Interferometer

As demonstrated in the previous experiments using discrete components, maintaining phase and polarization stability is essential for effective interferometric recombination. A monolithically integrated OEO interferometer photonic integrated circuit (PIC) offers significant advantages in enhancing polarization and phase stability. The indium phosphide (InP) platform, known for its high integration capabilities of active components [41, 42], is an excellent choice for developing a monolithically integrated OEO interferometer PIC. Two generations of the InP-based integrated OEO interferometer are introduced. Initially, a partially integrated OEO interferometer is presented. This PIC incorporates waveguide routing for the interferometer setup and a coherent photodetector. After characterizing the PIC, it is assembled into an OEO interferometer module. Subsequently, a fully integrated OEO interferometer PIC is introduced which is monolithically integrating all the active optical components necessary for the OEO interferometer setup.

2.4.1 Partially Integrated OEO Interferometer PIC

In this section the partially integrated OEO interferometer is presented. The partially integrated PIC essentially combines the waveguide routing for the OEO interferometer and a coherent photodetector. It is an intermediate step towards the fully integrated OEO interferometer. The PIC is assembled into an OEO interferometer module.

2.4.1.1 PIC Design and Fabrication

The partially integrated OEO interferometer PIC (Fig. 2.23) was manufactured on the same InP platform as in [43]. This PIC integrates a coherent photodetector and waveguide routing for the OEO interferometer setup. The coherent photodetector is based on a 2 × 4 multimode interference coupler and four photodiodes. Due to the DSP delay compensation necessitating an optical delay line of several meters, the reference path is implemented using off-chip fiber. To account for phase shifts within the off-chip reference path, a thermal phase shifter is placed

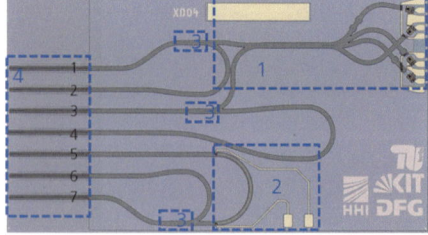

Fig. 2.23 Top-view photograph of the partially integrated OEO interferometer PIC: (1) coherent photodetector; (2) heater; (3) couplers; (4) spot-size converters

on the chip. For efficient fiber chip coupling, the processed chip comprises single-mode-fiber spot-size converters as optical interfaces.

2.4.1.2 PIC Measurement

Figure 2.24a shows the responsivity measurement of the coherent photodetector on the chip for the entire C- and L-band prior to assembly. It shows the results for TE polarized input light. The dashed lines refer to measurements using optical interface 1 whereas the other refer to measurements using optical interface 3 (Fig. 2.23). The results indicate that the coherent photodetector's center wavelength is approximately 1550 nm, as intended, with a maximum responsivity of 0.061 A/W for input WG1 and 0.055 A/W for input WG3. These values account for a 3 dB inherent splitter loss. The lower maximum responsivity for WG3 compared to WG1 is due to additional waveguide bends. Figure 2.24b displays a dark current measurement of the photodiodes, showing a dark current of 1 nA at a reverse bias voltage of 2 V.

2.4.1.3 Module Setup

The assembled module set-up consists of a housing positioned on an evaluation board (Fig. 2.25). For packaging, a generic housing from the EU project PIXAPP was utilized. Inside the housing, an eight-channel fiber array is optically coupled to the chip's facets. The interferometer chip and a transimpedance amplifier (TIA) are mounted on a ceramic substrate, with the TIA electrically connected to the coherent receiver pads via wire bonds. The DC and RF pads of the TIA are electrically linked to the corresponding connections within the housing. In an OEO interferometer setup, the module functions both as the receiver, converting the optical signal into the electrical domain, and as the foundation of the optical network, facilitating optical connections among all components of the OEO interferometer.

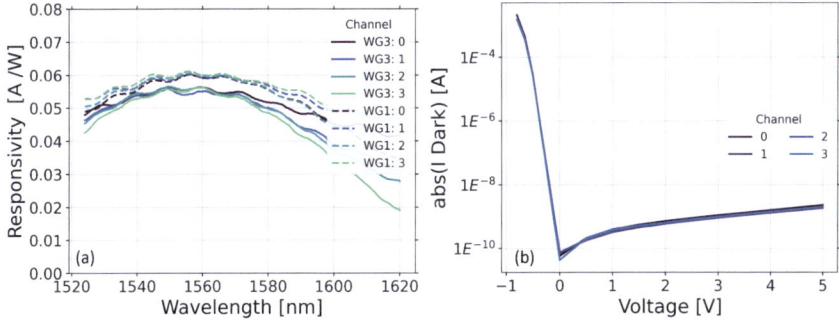

Fig. 2.24 **a** Responsivity and **b** dark current measurement of the coherent photodetector

Fig. 2.25 Photograph of the assembled module with its evaluation board, corresponding to the chip schematic shown in Fig. 2.23

2.4.1.4 Module Measurement

After assembling, a heterodyne bandwidth measurement of the module was conducted, using an electrical spectrum analyzer (ESA) (Fig. 2.26). The results presented are for TE polarized input light at the wavelength of 1550 nm. As illustrated in Fig. 2.26, the results of channel 1 to 3 are consistent while channel 2 shows a different behavior for the measured frequencies.

2.4.2 Fully Integrated OEO Interferomter PIC

In this section the fully integrated OEO interferometer PIC is presented. It comprises all optical components necessary for an OEO interferometer setup.

2.4.2.1 PIC Design and Fabrication

The fully integrated OEO interferometer PIC is fabricated on an indium phosphide platform, essentially identical to the Fraunhofer Heinrich Hertz Institute multi-project wafer InP foundry platform [42]. Figure 2.27 shows a top-view photograph of the presented PIC. The PIC integrates a coherent photodetector, a tunable local oscillator (LO), an IQ modulator,

Fig. 2.26 Bandwidth measurement of all four channels of the assembled module

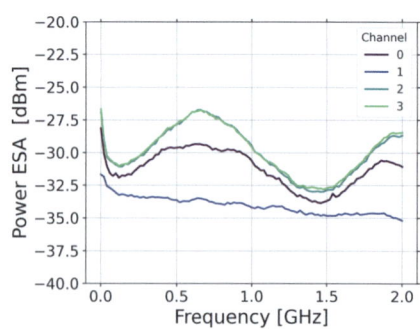

2 Optic-Electronic-Optic Interferometers for Ultrabroadband …

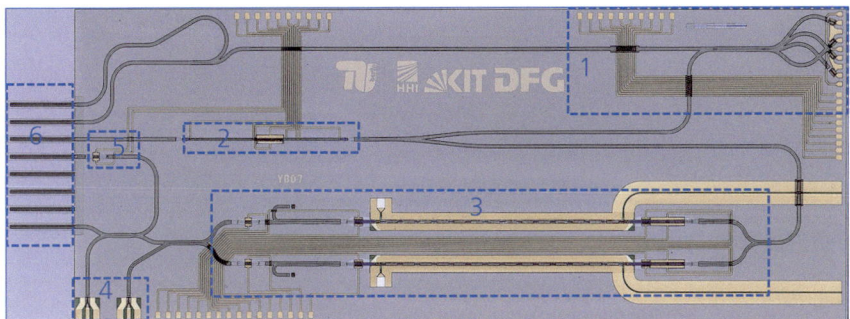

Fig. 2.27 Photograph of fully integrated OEO interferometer PIC: (1) coherent photodetector; (2) tunable LO; (3) IQ MZM; (4) reference photodiodes; (5) EO phase shifter; (6) spotsize converters

passive waveguide routing, reference photodiodes for measurement and spotsize converters for efficient fiber chip coupling. As for the partially integrated PIC, the delay line of the reference path is not integrated into the PIC due to its expected length of several meters. In [44], a novel concept for an off-chip delay line with inherent phase stability for the OEO interferometer is introduced. However, this concept is incompatible with an integrated LO, and consequently, it is not utilized in the fully integrated OEO interferometer PIC. An electro-optic (EO) phase shifter is integrated onto the chip to compensate for phase shifts occurring within the off-chip delay line. The coherent photodetector is based on a 2×4 multimode interference coupler and four photodiodes [43]. For the tunable LO a sampled grating distributed Bragg reflector (SGDBR) is used. It is designed according to [45]. The IQ modulator comprises two traveling wave electrode (TWE)-based Mach-Zehnder phase modulators (MZMs), each with a TWE length of 3.5 mm [46].

2.4.2.2 PIC Measurement

The coherent photodetector of the PIC is characterized by measuring the responsivity and the dark current of all four photodiodes. Figure 2.28a shows the measured responsivity of all four photodiodes for TE polarized light. The measurement was performed by using an external laser source and a reverse bias voltage of 2 V. The maximum is at 0.055 A/W at a center wavelength of 1580 nm. The value includes 3 dB inherent splitter loss. The center wavelength indicates that the photodetector is centered within the C- and L-bands. Figure 2.28b illustrates the dark current measurement of the photodiodes. At a reverse bias voltage of 2 V, the photodiodes exhibit a dark current of 10 nA.

Figure 2.29 shows the LO tuning range. The LO is tunable over a wavelength range of 43 nm, ranging from 1551 nm to 1594 nm. Without tuning the center wavelength of the LO is at 1563 nm.

The two MZMs of the IQ modulator are DC characterized by measuring both V_π and the extinction ratio (ER). This measurement utilizes the untuned on-chip local oscillator as the

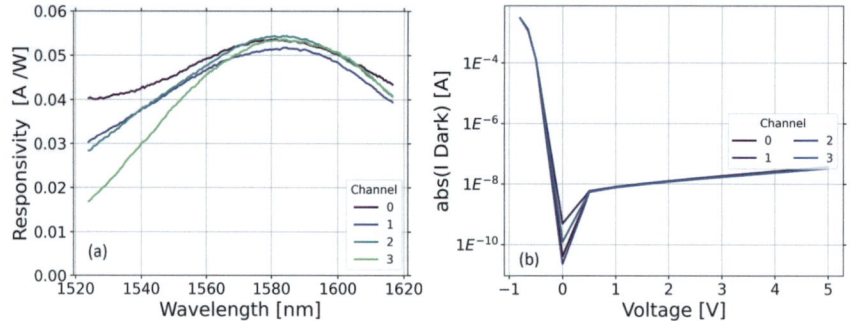

Fig. 2.28 a Responsivity and b dark current measurement of the coherent photodetector

Fig. 2.29 Tuning range of monolithically integrated LO

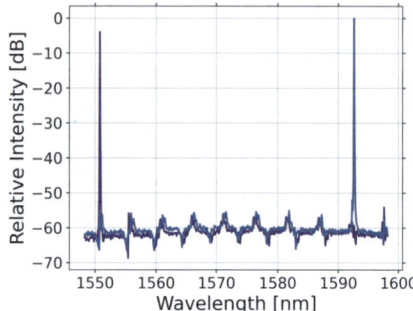

light source, with reverse bias voltages applied to both traveling wave electrode (TWE) electrodes (P1 and P2). The output power of the MZMs is measured, using one of the integrated reference photodiodes. Figure 2.30 presents the extinction ratio (ER) measurements for both MZMs at 1563 nm, indicating that the maximum ER for each exceeds 10 dB. The ER is constrained by two reasons: firstly, light propagating as substrate modes will be detected by the integrated photodiode. Second, while characterizing one MZM the other must be turned off by putting it into its off-state. Residual light passing through the MZM in the off-state will interfere with the signal to be measured, thus limiting the ER measurement. From Fig. 2.30 for the upper MZM V_π can be deduced to be 4.3 V at a bias at of −5.5 V. And for the lower MZM can be read as 6.2 V with a bias of −5 V. The 3 dB RF bandwidth of both the MZMs and the photodiodes is expected to be approximately 25 GHz.

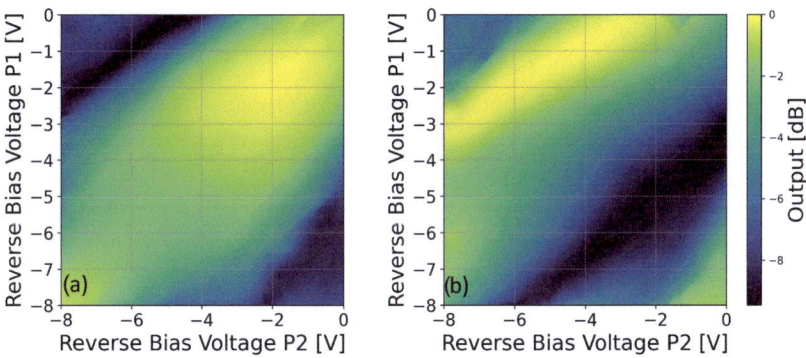

Fig. 2.30 Extinction ratio and V_π measurement of the **a** upper and **b** lower MZM of the IQ-modulator

2.4.2.3 Module Setup

In upcoming OEO interferometer demonstrations featuring the monolithically integrated OEO interferometer PIC, the plan is to integrate the PIC with a TIA into an OEO interferometer module. The packaging will be similar to that used for the partially integrated version. Figure 2.31 displays a 3D model of the intended OEO interferometer module.

Fig. 2.31 Model of planned fully integrated OEO interferometer module

2.5 Summary

We present the first experimental demonstration of an OEO interferometer performing coherent optical add-drop functionality. The system development is presented methodically, beginning with a coherent heterodyne detection approach for single-channel input and advancing to an intradyne detection configuration suitable for WDM signals. A comprehensive framework for the required DSP algorithms is established, along with a discussion of critical design aspects such as optical path length matching and phase stabilization. In the current implementation, the achievable symbol rate is primarily limited by the sampling rates of the RFSoC's data converters. Advancements in FPGA technology, along with faster, integrated, and synchronous data converters, are expected to enable higher symbol rates in future implementations.

The transition to higher-order modulation formats is mainly constrained by the achievable extinction ratio of the destructive interference process, which remains a critical performance factor. Extending the system to support overlapping subcarrier multiplexed signals, such as multicarrier offset quadrature amplitude modulation (OQAM) requires symbol-level signal recovery with complete frequency and phase estimation, as well as reapplication at the remodulation stage.

System stability is currently affected by phase and polarization drifts, largely due to the use of discrete optical components. This limitation is addressed through photonic integration. To this end, two generations of InP-based integrated OEO interferometer PICs are presented. As an initial step, a partially integrated OEO interferometer PIC is designed, characterized, and assembled into a module, with experimental validation confirming the feasibility of further integration. This is followed by the realization and characterization of a fully integrated OEO interferometer PIC. Current efforts focus on packaging the fully integrated PIC into a module and experimentally validating its performance and robustness. These developments contribute to the realization of fully integrated, high-performance photonic systems for advanced coherent optical networking, computing, and sensing.

References

1. Drayss D, Fang D, Sherifaj A, Peng H, Füllner C, Henauer T, Lihachev G, Freude W, Randel S, Kippenberg T, Zwick T, Koos C (2023) CLEO 2023. Optica Publishing Group, p STh5C.8. https://doi.org/10.1364/CLEO_SI.2023.STh5C.8
2. Füllner C, Sherifaj A, Henauer T, Fang D, Drayss D, Harter T, Gutema TZ, Freude W, Randel S, Koos C (2022) Conference on lasers and electro-optics. Optica Publishing Group, p STh5M.3. https://doi.org/10.1364/CLEO_SI.2022.STh5M.3
3. Singh K, Meier J, Misra A, Preußler S, Scheytt JC, Schneider T (2020) IEEE Photonics Technol Lett 32(24):1544. https://doi.org/10.1109/LPT.2020.3039621

4. Moscoso-Mártir A, Schulz O, Misra A, Merget F, Pachnicke S, Witzens J (2023) Opt Commun 546:129809. https://doi.org/10.1016/j.optcom.2023.129809
5. Koos C, Leuthold J, Freude W, Kohl M, Dalton L, Bogaerts W, Giesecke AL, Lauermann M, Melikyan A, Koeber S, Wolf S, Weimann C, Muehlbrandt S, Koehnle K, Pfeifle J, Hartmann W, Kutuvantavida Y, Ummethala S, Palmer R, Korn D, Alloatti L, Schindler PC, Elder DL, Wahlbrink T, Bolten J (2016) J Light Technol 34(2):256. https://doi.org/10.1109/JLT.2015.2499763
6. Gnauck A, Winzer P (2005) J Lightwave Technol 23(1):115. https://doi.org/10.1109/JLT.2004.840357
7. Marom DM, Colbourne PD, D'Errico A, Fontaine NK, Ikuma Y, Proietti R, Zong L, Rivas-Moscoso JM, Tomkos I (2017) J Opt Commun Netw 9(1):1. https://doi.org/10.1364/JOCN.9.000001
8. McMahon PL, Marandi A, Haribara Y, Hamerly R, Langrock C, Tamate S, Inagaki T, Takesue H, Utsunomiya S, Aihara K, Byer RL, Fejer MM, Mabuchi H, Yamamoto Y (2016) Science 354(6312):614. https://doi.org/10.1126/science.aah5178
9. Fabbri SJ, Sygletos S, Perentos A, Pincemin E, Sugden K, Ellis AD (2015) J Light Technol 33(7):1351. https://doi.org/10.1109/JLT.2015.2390292
10. Richter T, Schmidt-Langhorst C, Elschner R, Kato T, Tanimura T, Watanabe S, Schubert C (2015) J Light Technol 33(3):685. https://doi.org/10.1109/JLT.2014.2379472
11. Taylor MG (2011) Coherent optical channel substitution. US Patent 8,050,564B2, 1 Nov 2011
12. Schell M, Vorreau P (2015) Network element. US Patent 8,971,726B2, 3 Mar 2015
13. Winzer PJ (2013) J Light Technol 31(11):1775. https://doi.org/10.1109/JLT.2013.2257687
14. Mahmud MS, Kemal JN, Adib M, Fullner C, Schindler A, Runge P, Schell M, Freude W, Koos C, Randel S (2020) Conference on lasers and electro-optics. Optica Publishing Group, p SF1L.1. https://doi.org/10.1364/CLEO_SI.2020.SF1L.1
15. Tempus M, Lüthy W, Weber HP (1993) Appl Phys B Photophys Laser Chem 56(2):79. https://doi.org/10.1007/BF00325244
16. Freschi AA, Frejlich J (1995) Opt Lett 20(6):635. https://doi.org/10.1364/ol.20.000635
17. Mahmud MS, Matalla P, Adib MMH, Koos C, Randel S (2023) CLEO 2023. Optica Publishing Group, p SM2I.6. https://doi.org/10.1364/CLEO_SI.2023.SM2I.6
18. Mahmud MS, Matalla P, Dittmer J, Koos C, Randel S (2025) Opt Express 33(4):6885. https://doi.org/10.1364/OE.532854
19. Lyons RG (2004) Understanding digital signal processing, 2nd edn. Prentice Hall PTR, Chap, p 9
20. Füllner C, Adib MMH, Wolf S, Kemal JN, Freude W, Koos C, Randel S (2019) J Light Technol 37(17):4295. https://doi.org/10.1109/JLT.2019.2923249
21. Leven A, Kaneda N, Koc UV, Chen YK (2007) IEEE Photonics Technol Lett 19(6):366. https://doi.org/10.1109/LPT.2007.891893
22. Li L, Tao Z, Oda S, Hoshida T, Rasmussen JC (2008) OFC/NFOEC 2008–2008 conference on optical fiber communication/national fiber optic engineers conference, pp 1–3. https://doi.org/10.1109/OFC.2008.4528776
23. Hoffmann S, Bhandare S, Pfau T, Adamczyk O, Wordehoff C, Peveling R, Porrmann M, Noe R (2008) IEEE Photonics Technol Lett 20(18):1569
24. Nakashima H, Tanimura T, Hoshida T, Oda S, Rasmussen JC, Li L, Tao Z, Ishii Y, Shiota K, Sugitani K, Adachi H (2008) 2008 34th European conference on optical communication, pp 1–2. https://doi.org/10.1109/ECOC.2008.4729128
25. Matalla P, Mahmud MS, Füllner C, Koos C, Freude W, Randel S (2021) 2021 optical fiber communications conference and exhibition (OFC), pp 1–3
26. Savory SJ (2008) Opt Express 16(2):804. https://doi.org/10.1364/OE.16.000804
27. Kuschnerov M, Hauske FN, Piyawanno K, Spinnler B, Alfiad MS, Napoli A, Lankl B (2009) J Light Technol 27(16):3614. https://doi.org/10.1109/JLT.2009.2024963

28. Viterbi AJ, Viterbi AM (1983) IEEE Trans Inf Theory 29(4):543. https://doi.org/10.1109/TIT.1983.1056713
29. Herzog F, Kudielka K, Erni D, Bächtold W (2006) IEEE J Quantum Electron 42(10):973. https://doi.org/10.1109/JQE.2006.881413
30. Kudielka K, Klaus W (1999) 1999 IEEE LEOS annual meeting conference proceedings. LEOS'99. 12th annual meeting. IEEE lasers and electro-optics society 1999 annual meeting (Cat. No. 99CH37009), vol 1, pp 295–296. doi: 10.1109/LEOS.1999.813599
31. Ferrero V, Camatel S (2008) Optic Express 16(2):686
32. Ashok R, Naaz S, Kamran R, Sidhique A, Gupta S (2020) 2020 IEEE photonics conference, IPC 2020–proceedings, pp 31–32. https://doi.org/10.1109/IPC47351.2020.9252215
33. Gardner F (2005) Phaselock techniques. Wiley
34. Ziegler JG, Nichols NB (1993) J Dyn Syst Meas Control 115(2B):220. https://doi.org/10.1115/1.2899060
35. Personick SD (1971) Bell Syst Tech J 50(1):213. https://doi.org/10.1002/j.1538-7305.1971.tb02544.x
36. Kazovsky LG (1985) J Opt Commun 6(1):18. https://doi.org/10.1515/JOC.1985.6.1.18
37. Barry JR, Lee EA (1990) Proc IEEE 78(8):1369. https://doi.org/10.1109/5.58322
38. Ip E (2018) Optical coherent detection and digital signal processing of channel impairments. Springer, Singapore, pp 1–70
39. Mahmud MS, Matalla P, Dittmer J, Schindler A, Runge P, Koos C, Randel S (2024) CLEO 2024. Optica Publishing Group, p STh1Q.6. https://doi.org/10.1364/CLEO_SI.2024.STh1Q.6
40. Chaudhari Q (2018) Wireless communications from the ground up: an SDR perspective. CreateSpace Independent Publishing Platform
41. Arafin S, Coldren LA (2018) IEEE J Sel Top Quantum Electron 24(1):1. https://doi.org/10.1109/JSTQE.2017.2754583
42. Soares FM, Baier M, Gaertner T, Grote N, Moehrle M, Beckerwerth T, Runge P, Schell M (2019) Appl Sci 9(8). https://doi.org/10.3390/app9081588. https://www.mdpi.com/2076-3417/9/8/1588
43. Runge P, Schubert S, Seeger A, Janiak K, Stephan J, Trommer D, Nielsen ML (2014) IEEE Photonics Technol Lett 26(4):349. https://doi.org/10.1109/LPT.2013.2293635
44. Sartorius B, Runge P, Schell M (2020) Optische anordnung und verfahren zur optischen signalverarbeitung. Patent number: DE20191020641520190503
45. Lee MH, Soares F, Baier M, Möhrle M, Rehbein W, Schell M (2020) Semiconductor lasers and laser dynamics IX. In: Sciamanna M, Michalzik R, Panajotov K, Höfling S (eds) International society for optics and photonics (SPIE), vol 11356, p 1135605. doi: 10.1117/12.2554541
46. Gupta YD, Binet G, Diels W, Abdeen O, Gaertner T, Baier M, Schell M (2023) J Light Technol 41(11):3498. https://doi.org/10.1109/JLT.2023.3244129

Open Access This chapter is licensed under the terms of the Creative Commons Attribution 4.0 International License (http://creativecommons.org/licenses/by/4.0/), which permits use, sharing, adaptation, distribution and reproduction in any medium or format, as long as you give appropriate credit to the original author(s) and the source, provide a link to the Creative Commons license and indicate if changes were made.

The images or other third party material in this chapter are included in the chapter's Creative Commons license, unless indicated otherwise in a credit line to the material. If material is not included in the chapter's Creative Commons license and your intended use is not permitted by statutory regulation or exceeds the permitted use, you will need to obtain permission directly from the copyright holder.

Ultra-Broadband Photonically Assisted Analog-to-Digital-Converters

Jeremy Witzens, Daniel Drayss, Dengyang Fang, Alvaro Moscoso Mártir, Juliana Müller, Maxim Weizel, Andrea Zazzi, Wolfgang Freude, Christian Koos, Sebastian Randel and J. Christoph Scheytt

Abstract

We present recent progress made towards ultra-broadband photonically assisted analog-to-digital converters, that leverage both the low jitter of best-of-class mode-locked lasers as well as the capability of optics to break down broadband signals into multiple lower speed tributaries that can be better handled by electronics. We review in particular our work on both time- and frequency-domain approaches and give an outlook on how these architectures can be extended to include further signal processing tasks such as equalization. Optically triggered track-and-hold amplifiers are reported with an equivalent jitter below 80 fs rms in a signal frequency range from 20 GHz to 70 GHz. Frequency-domain architectures implementing optical arbitrary waveform measurement up to signal bandwidths of 610 GHz are also shown. Finally, an architecture allowing the deserialization and equalization of PAM4 signals is introduced and modeled for operation in 400 Gb/s links.

J. Witzens (✉) · A. Moscoso Mártir
Institute of Integrated Photonics, RWTH Aachen University, Aachen, Germany
e-mail: jwitzens@iph.rwth-aachen.de

D. Drayss · D. Fang · W. Freude · C. Koos · S. Randel
Institute of Photonics and Quantum Electronics, Karlsruhe Institute of Technology, Karlsruhe, Germany

J. Müller
Now at Black Semiconductor, Aachen, Germany

M. Weizel · J. C. Scheytt
Heinz Nixdorf Institute, Paderborn University, Paderborn, Germany

A. Zazzi
Now at University of California at Berkley, Berkley, CA, USA

© The Authors(s) 2026
J. C. Scheytt et al. (eds.), *Electronic-Photonic Integrated Systems for Ultrafast Signal Processing*, https://doi.org/10.1007/978-3-032-08340-1_3

Acronyms

ADC	Analog-to-Digital Converter
AWG	Arrayed Waveguide Grating
AWGN	Additive White Gaussian Noise
BOMPD	Balanced Optical Microwave Phase Detector
BPD	Balanced Photodetector
BW	Bandwidth
C-RAN	Cloud Radio-Access Network
CROW	Coupled-Resonator Optical Waveguide
CSNR	Constellation Signal-to-Noise Ratio
DC	Directional Coupler
DCS	Directional Coupler Splitter
DSP	Digital Signal Processing
ECL	External-Cavity Laser
EDFA	Erbium-Doped Fiber Amplifier
EM	Electromagnetic
ENOB	Effective Number of Bits
EO	Electro-Optic
EPIC	Electronic-Photonic Integrated Circuit
FBG	Fiber Bragg Grating
FDE	Finite Difference Eigenmode
FSR	Free Spectral Range
FDTD	Finite Difference Time Domain
FWHM	Full Width at Half Maximum
GC	Grating Coupler
HF	High-Frequency
IC	Integrated Circuit
IQR	In-Phase and Quadrature Receiver
ISI	Inter-Symbol Interference
LO	Local Oscillator
MFD	Mode-Field-Diameter
MLL	Mode-Locked Laser
MMI	Multimode Interference
MRC	Maximum-Ratio Combining
MZM	Mach-Zehnder Modulator
OADM	Optical Add-Drop Multiplexer
OAWM	Optical Arbitrary-Waveform Measurement
OCNR	Optical Carrier-to-Noise Ratio
ODDM	Orthogonal Delay Division Multiplexing
OEPLL	Opto-Electronic Phase Locked Loop

OFC	Optical Frequency Comb
OH	Optical Hybrid
OR	Overlap Region
ORR	Optical Ring Resonator
PAM	Pulse Amplitude Modulation
PAM4	4-Level Pulse Amplitude Modulation
PCB	Printed Circuit Board
PIC	Photonic Integrated Circuit
PSD	Power Spectral Density
QAM	Quadrature Amplitude Modulation
RF	Radio Frequency
SBVT	Sliceable Bandwidth-Variable Transponder
SEF	Switched-Emitter-Follower
SI	Supplementary Information
SINAD	Signal-to-Noise-and-Distortion Ratio
SNDR	Signal-to-Noise-and-Distortion Ratio
SNR	Signal-to-Noise Ratio
SSMF	Standard Single Mode Fiber
TFLN	Thin-Film Lithium-Niobate
THA	Track-and-Hold Amplifier
TIA	Transimpedance Amplifier
VCO	Voltage Controlled Oscillator
WDM	Wavelength-Division Multiplexing
WS	Waveshaper-Programmable Optical Filter
YIG	Yttrium Iron Garnet

3.1 Introduction

The sampling rate, analog bandwidth and effective number of bits (ENOB) of analog-to-digital converters (ADCs) are becoming a bottleneck for the scaling of data rates in fiber optic communications. While ADCs have been an essential component of long-haul fiber optic communication links for several decades, the introduction of higher-order pulse amplitude modulation (PAM) and of asynchronous digital signal processing in intra-datacenter interconnects has also moved them into focus there [1]. The ENOB of ADCs in particular is limited by the jitter of electrical clocks, as generated by quartz oscillators in conventional systems. This has motivated the use of mode-locked lasers (MLLs) as low jitter frequency references in analog-to-digital conversion [2], since benchtop MLLs have been shown to provide pulse trains with jitter that outperforms that of quartz oscillators by many orders of magnitude [3].

This chapter deals with progress on optically enabled ADC architectures developed with the goals of leveraging the low jitter of such optical clocks and to reach analog bandwidths and sampling rates outperforming those of pure electronic systems. These are broadly categorized into time- and frequency-domain architectures, wherein the pulse train of the MLL acts as a time reference in the former and the comb-like spectrum of the MLL acts as a frequency reference in the latter. Notably, even for the frequency-domain architectures, the jitter of the pulse train remains a primary driver of the overall system performance after signal aggregation [4].

Section 3.2 deals with optically-triggered electrical track-and-hold amplifiers (THAs), in which optical pulses replace an electrical clock transition. In Sect. 3.2.1, the THA is directly triggered by the optical pulse, with the simplicity of this approach enabling high analog bandwidth and low jitter. As a disadvantage, this requires the optical pulse train to already have the required repetition rate, which can be problematic as best-of-class MLLs are typically benchtop devices with large cavities and consequently large repetition times. Since in this architecture only the envelope function of the pulse train matters and not the actual free spectral range (FSR) in the frequency domain, the effective repetition rate of the pulses can be increased with low optical losses with a pulse interleaver [5]. Alternatively, a higher repetition rate voltage controlled oscillator (VCO) can be locked to a multiple of the repetition rate of the MLL with an opto-electronic phase locked loop (OEPLL), as described in Sect. 3.2.2.

In Sect. 3.3, two approaches are described in which an optical front-end converts an incoming broadband optical signal into several lower-speed tributaries, each of which can be digitized with downstream all-electrical ADCs whose bandwidth and sampling rate requirements are correspondingly reduced. In both architectures, signal reconstruction is done in the frequency domain, wherein the first relies in first slicing the optical signal in the frequency domain prior to digitization and stitching, while the second relies on coherent detection of the signal using a frequency comb generated by an MLL as a local oscillator, with different group delays applied to the comb for each tributary. This second method has the advantage of not requiring optical filters and thus facilitates the configuration of the chip, as higher-order optical filters are not required. Such filters are used in the first, frequency sliced architecture and typically require several phase tuners each to compensate for fabrication errors and temperature drift. While both methods rely on frequency-domain signal processing and do not require narrow pulses, i.e., the optical comb can be dispersed, the second one can be reduced to coherent sampling of the signal by time-interleaved pulses in the undispersed case, and can thus also be considered as an intermediate case between the time- and frequency-domain approaches.

A signal multiplexing approach relying on generating orthogonal carriers by delaying an optical comb by different group delays (orthogonal delay division multiplexing—ODDM) [6] can be reduced to an optical deserializer with equivalent functionality [7], that can be further generalized to implement equalization in the optical domain prior to digitization, further reducing the ENOB requirement of the utilized electrical ADCs. Here too, a filter-less approach facilitates the configuration of the system and increases its tolerance against temperature drift [6].

3.2 Optically Triggered Track-and-Hold Amplifiers as Analog Front-Ends for Time-Interleaved ADCs

Time-interleaved ADCs are used in a variety of modern digital data acquisition systems, such as next-generation electronic/optic communication systems, high-performance measurement and terahertz signal processing. By parallelization over N channels, the aggregate sampling rate f_s can be scaled up by the interleaving factor N. Unfortunately, the bandwidth (BW) does not scale this way and each channel needs to support the full system bandwidth. The sampling-rate of downstream tributary channels, on the other hand, has to be greater than $2 \cdot \text{BW}/N$. To maintain a high bandwidth in the range of tens of GHz and beyond, THAs are used as analog ADC sampling front-ends with a moderate interleaving factor of two to four. With increasing system bandwidth and thus sampling rate requirements, generating N precisely time-interleaved clock signals with sufficiently low phase noise/jitter is a major challenge. A well-known and often used formula that describes the effect of timing jitter onto the achievable signal-to-noise ratio (SNR) of a data converter is given by

$$\text{SNR}_\text{dB} = -20 \cdot \log_{10}\left(2\pi f_\text{in} \cdot \sigma_\text{ji}\right) \quad (3.1)$$

where f_in is the input frequency and σ_ji the standard deviation of the timing jitter [8]. The equation is derived under the assumption that $2\pi f_\text{in}\sigma_\text{ji} \ll 1$ and exemplifies that the SNR decreases with increasing timing-jitter variance and input frequency. Advancements in ultra-low timing jitter femtosecond MLLs, with which attosecond-precision distribution of optical pulse trains has been demonstrated [9], provide a path to achieve high SNR at high input frequencies by shifting the clock generation, manipulation and distribution into the optical domain.

In the following, we describe two approaches to synchronize time-interleaved THAs with an optical pulse train, the first relying on direct optical triggering and the second on locking a VCO to it.

3.2.1 Track-and-Hold Amplifiers with Direct Optical Triggering

In a directly optically-triggered THA, the power transported by the optical signal defines the clock signal and thus the sampling time. In the envisioned time-interleaved architecture, the light is distributed by means of optical waveguides to N individual THA channels with channel-dependent time-delay increments that are multiples of T_s/N, with T_s the repetition period of the pulse train, see Fig. 3.1. The optical signal triggers the sampling process via e.g.

photo-conductive switches or is converted into electrical currents by means of photodiodes that are further processed by passive or active transimpedance amplifiers (TIAs).

3.2.1.1 Concepts

The development of suitable on-chip MLLs is an active area of research, however, best-of-class MLLs currently remain solitary devices [10], so that we envision a partial integration as represented in Fig. 3.1. The light distribution and delay network is implemented on chip together with the BiCMOS electronics in a silicon electronic-photonic integrated circuit (EPIC) platform. Only the comb generation and spectral shaping are implemented off chip. While progress has been made in reducing the RF linewidth and jitter of semiconductor MLLs by combining III-V gain media with silicon photonic integrated circuits (PICs), either in extended cavity [11] or in external feedback configuration [12], the limited gain spectrum of semiconductors and the resulting large pulse durations present a fundamental limit to reducing the jitter [13]. Promising work on the chip-scale integration of optically pumped rare-earth MLLs with a sufficiently wide gain spectrum is, however, in progress [14] and reported in another chapter of this book.

Optical Clock Signal Generation: We choose an optical clock configuration in which high power encodes a logical high and sets the THA into track-mode and low optical power (or off) encodes a logical low and switches the circuit into hold-mode. This translates short MLL pulses into a sharply defined sampling time and contributes to maintaining a low jitter. As mentioned above, Eq. (3.1) shows that with increasing analog input frequency the effect of clock jitter on the SNR also increases and becomes the limiting factor. This is the motivation behind the use of ultra-low phase-noise MLLs.

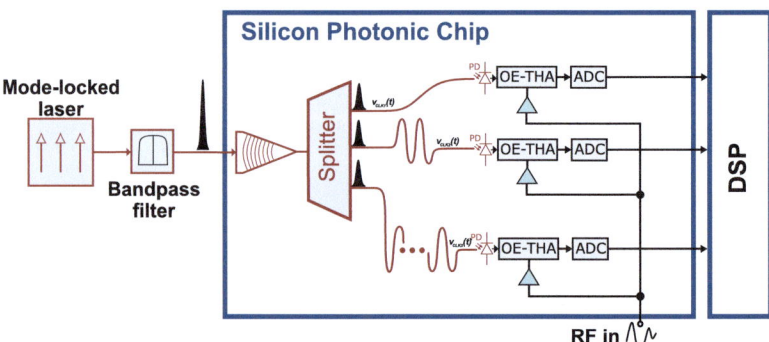

Fig. 3.1 Time-interleaved optically-clocked/triggered ADC concept in a silicon photonics technology

Phase-noise can be converted into the system jitter using [15, p. 54]

$$\sigma_{ji} = \frac{1}{2\pi f_0} \sqrt{2 \int_{f_{min}}^{\frac{f_0}{2}} \mathcal{L}(f)\, df} \tag{3.2}$$

where $\mathcal{L}(f)$ is the dual-sided power spectral density (PSD) of the phase-noise and f_0 is both the fundamental frequency of the clock signal and the sampling rate of the THA. The lower integration limit f_{min} is determined by the observation time, which is limited in practical applications [15, p. 54]. $f_0/2$ can be chosen as an upper limit, since it corresponds to the Nyquist frequency and thus to the required bandwidth.

Clock Distribution: The MLL pulse train has to be split and distributed to the individual THA/ADC channels, for example with discrete fibers or on-chip silicon or silicon-nitride waveguides. In Fig. 3.1 the pulse train is first bandpass-filtered and then coupled into the EPIC by means of a grating coupler (GC). The light is then split into N separate paths with optical delay lines that generate the necessary time-delays for interleaving the N clock signals. In Fig. 3.1, a simple power splitter is shown as this maintains the pulse shape. This differs from the wavelength demultiplexing approach used in [2, 16], that employs an arrayed waveguide grating (AWG) and coupled resonator optical waveguide (CROW) filters, respectively, to generate interleaved pulses intended to carry the data.

Detection and Amplification (Opto-Electronic Conversion): Each channel consists of an opto-electronic/optically-triggered THA. A photodiode converts the optical signal into an electrical current and a passive (resistive) or active TIA with subsequent clock signal buffer is used to drive the electronic switches. For a high-speed sampler with sampling rates in the range of tens of GHz, an active TIA is better suited. It allows for a more precise control over the gain, bandwidth and linearity requirements.

Sampling Electronics and ADCs: High-speed THAs in SiGe or InP can be used as time-interleaved analog ADC front-ends. Subsequently, the electronic input signal (RF_{in}) is resampled by slower CMOS ADCs and converted into its digital representation. The resolution of the ADCs needs to be at least as high as the targeted ENOB. Using a BiCMOS EPIC platform, they can be combined on a single chip with the silicon photonics front-end as represented in Fig. 3.1. A single channel of such an optically-triggered THA is designed in a silicon photonics SiGe BiCMOS EPIC technology and analyzed by means of simulations in [17]. The results indicate a small-signal bandwidth of 78 GHz and over 7 bit effective resolution up to a 50 GHz signal bandwidth. To verify the high performance shown in these simulations, the chip was fabricated and is described in the next section.

3.2.1.2 Demonstration

Using the IHP SG25H5 EPIC SiGe BiCMOS silicon photonics process, a one channel optically-triggered analog sampling front-end was demonstrated in [18] whose micrograph can be seen in Fig. 3.2a. The electronics are comprised of a THA in the switched-emitter-follower (SEF) topology and a common-base TIA for the opto-electronic clock conversion.

The SEF architecture is chosen because it typically achieves the highest possible bandwidth and sampling rate in a given technology. Advanced circuit techniques like emitter-degeneration with capacitive peaking and the inherent peaking behavior of a capacitively loaded emitter-follower are leveraged to achieve a ~65 GHz 3 dB small-signal bandwidth with a technology transit frequency f_T of 240 GHz.

In the photonic part of the EPIC, a GC is used to couple the light in. The rectangular shaped GC is optimized to receive a beam with a $1/e^2$ diameter of ~10 μm that matches the mode-field-diameter (MFD) of a cleaved standard single mode fiber (SSMF). A 300 μm long taper brings down the on-chip silicon-on-insulator waveguide (Si over SiO_2) to a width of 450 nm, at which it is single-mode. The light is further guided to a germanium p-i-n photodiode and transduced to a photocurrent, that is converted to a clock signal voltage by a subsequent TIA whose schematic is depicted in Fig. 3.2b. A common-base TIA topology is used mainly because of its simplicity and the possibility to control the photodiode reverse-bias voltage though the V_{pd_bias} terminal. It is very similar to a resistive TIA and the resistance $R_1 = 100\,\Omega$ defines the transimpedance, but it also allows for much higher bandwidth, since the photodiode only sees the input impedance looking into the emitter.

Simulations feature a TIA bandwidth of around 30 GHz. However, since at the time of the experiments only a low jitter MLL from Menhir Photonics with a central wavelength of 1550 nm and a repetition rate of 250 MHz was available, they were carried out with a 250 MHz sampling rate. The sparsity of samples resulting from operating in under-sampling mode at high signal frequencies does not, however, prevent the evaluation of the SNR and ENOB at high signal frequencies. The MLL generates an optical pulse train with an extremely

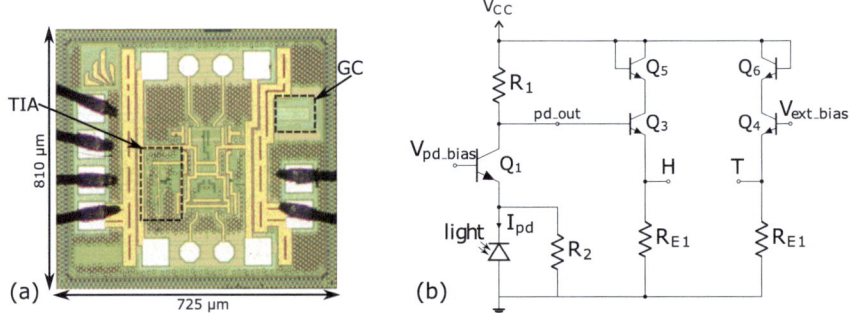

Fig. 3.2 a Micrograph of an opto-electronic THA. b Common-base TIA with simple output buffer. (a Reprinted with permission from [18] © Optical Society of America; b is taken from [19] used under CC-BY 4.0)

low jitter below 10 fs rms, integrated over an offset frequency range from 100 Hz to 10 MHz. The measured signal-to-noise-and-distortion (SINAD/SNDR) is plotted in Fig. 3.3b. Rearranging Eq. (3.1) for σ_{ji} results in a corresponding worst-case equivalent jitter below 80 fs rms in the signal frequency range from 20 to 70 GHz. The best equivalent jitter of 55.8 fs rms was obtained at 41 GHz. Figure 3.3a benchmarks this result against literature data, making apparent that it would be at the forefront of high-speed samplers if translated to a full ADC system.

3.2.2 Track-and-Hold Amplifiers Triggered by an Opto-Electronic Phase-Locked Loop

Unlike the optically-clocked/triggered THA described in the previous Sect. 3.2.1, in which photodiodes and the subsequent TIAs directly convert an MLL pulse train into an electronic clock, the OEPLL employs an MLL as an ultra-stable reference oscillator to stabilize an electronic oscillator that further drives the sampling switches. In particular, this enables up-conversion of the optical pulse train repetition rate.

3.2.2.1 Concepts

A conceptual schematic is shown in Fig. 3.4. The blue frame identifies a high-speed 2×-interleaved THA in a fast SiGe technology. A VCO, e.g. a Colpitts oscillator, is used to generate a clock signal in the range of 100 GHz. Placing it in between the two THAs and implementing it differentially inherently generates the 0° and 180° clock phases needed to drive the two time-interleaved sampling channels. Best-of-class bench-top MLLs usually

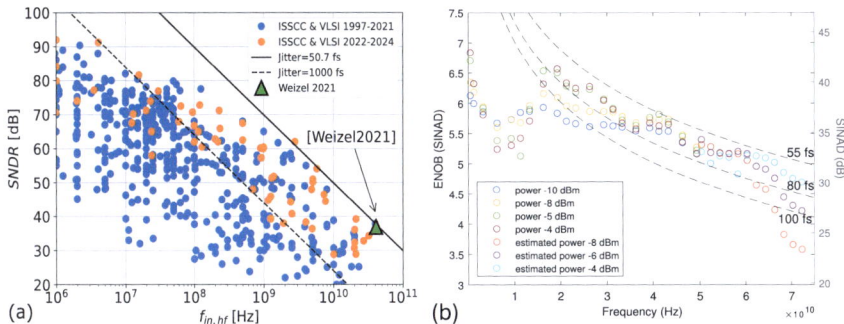

Fig. 3.3 a Measured SNDR vs. analog input frequency reported at the ISSCC and VLSI conferences from 1997 until 2024 [20]. The triangle represents the best-case measurement from [18]. **b** ENOB/SINAD measurements depending on signal frequency and amplitude. Dashed lines show the 55 fs, 80 fs and 100 fs equivalent jitter limits. (**b** is reprinted with permission from [18] © Optical Society of America)

do not have such high repetition rates, which can be addressed by incorporating a frequency divider in the OEPLL. A divide-by-four or divide-by-eight stage can be implemented on the high-frequency (HF) chip to reduce the chip-to-printed-circuit-board (PCB) or chip-to-chip interface requirements. An optional RF amplifier, which could also be already integrated on the HF side, drives a balanced optical microwave phase detector (BOMPD). This phase-detector can e.g. consist of a dual-output Mach-Zehnder modulator (MZM) as shown in Fig. 3.4. Lowering the frequency is beneficial because it allows the use of compact silicon photonic forward-biased phase-shifters inside the MZM [21]. After photo-detection, a low-pass filtered and amplified control voltage is generated to tune the on-chip VCO. As in Sect. 3.2.1, the MLL is kept as the only discrete element, due to the better phase-noise performance of bench-top devices. The advantage of this architecture is that the MLL repetition rate can be much lower than the required sampling rate. In fact, the PLL can not only lock to the fundamental frequency of the MLL, but also to any harmonic.

Here, we picture a hybrid system, using two chips or a chip + PCB combination. Although from a cost and compactness perspective it would be desirable to have a monolithically integrated solution, the available transistors in the EPIC platforms are not the fastest on the market. A hybrid two-chip approach allows to combine the best-of-class electronics with silicon photonics. HF-transistors are then implemented on a purely electronic integrated circuit (IC) on one side and low speed detection and control electronics for the OEPLL are integrated with the silicon photonics devices on the other.

The OEPLL concept is elaborated on in detail in [10, 22], where an yttrium iron garnet (YIG) oscillator with a tunable oscillation frequency of 2–20 GHz serves as the frequency-tuning element and a Menhir 1550 MLL with 250 MHz repetition rate serves as the reference. Phase noise measurements show an integrated rms jitter, integrated from 1 kHz to 100 MHz, of under 4 fs in the 5–20 GHz oscillation frequency range, with typical and minimum values of 4 fs and 3 fs, respectively.

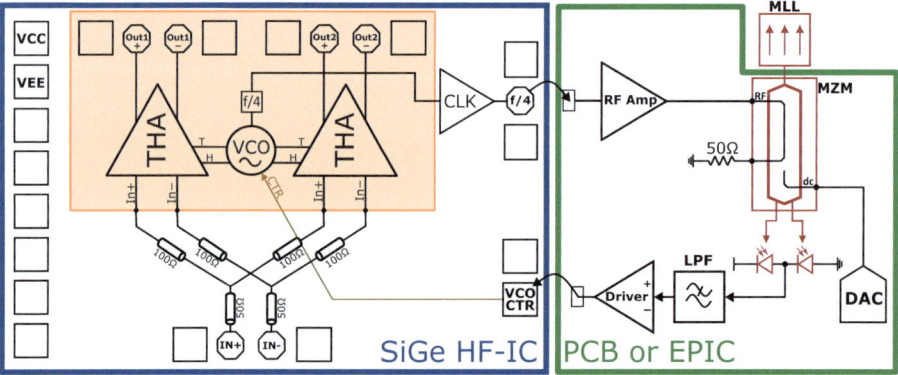

Fig. 3.4 Concept of a time-interleaved THA triggered by an opto-electronic PLL

3.2.2.2 Circuit Design and Implementation

To exemplify the reduction to practice, a high-speed THA is implemented in IHP's SG13G3 SiGe BiCMOS technology. It provides an f_T of 470 GHz, which is a significant improvement compared to the EPIC platform that has an f_T of only 240 GHz. Figure 3.5 shows the core-sampler schematic. A 2× time-interleaved architecture is chosen.

The input buffer is composed of two cascaded emitter-follower stages—first Q_1/Q_2, followed by Q_3/Q_4—with 100 Ω resistors (R_{IN}) providing input matching. After this stage, a switched pre-amplifier is used. It supplies a gain of 1 dB to offset losses in the signal chain and, to achieve a high bandwidth, several extension techniques are implemented. First, a cascode differential amplifier (Q_7-Q_{10}) is used to reduce the Miller effect, which results in a parasitic input capacitance of only $C_p = C_{be} + 2C_{bc}$, with C_{be} the base-emitter and C_{bc} the base-collector capacitance. In addition, a negative capacitance equal to $-C_{bc}$ is introduced by cross-coupling two transistors, effectively canceling out the base-collector capacitance of Q_7 and Q_8. The gain is finally tuned using resistor R_C and emitter degeneration R_E, while capacitive peaking with C_E extends the bandwidth without resorting to space-consuming inductors.

Transistors Q_{11} and Q_{12} are alternately switched. When Q_{11} is active, the circuit operates in track mode, allowing the input signal to appear at nodes V_{xp}/V_{xn}. When Q_{12} is active, these nodes are held at a constant voltage, thereby isolating the input during hold mode. The design then employs a pseudo-differential switched emitter-follower configuration as the core sampler. Only the positive side (Q_{19}) is shown in Fig. 3.5. This architecture offers both high bandwidth and fast sampling rates, making it ideal for analog ADC front-end applications. When Q_{19} is on, the voltage at the hold capacitor C_H follows the input. When it is off, the circuit enters hold mode.

Feed-forward capacitors (C_{ff}) further suppress feed-through, enhancing hold-mode isolation. A small resistor, R_{shunt}, is also added; although it slightly reduces overall bandwidth,

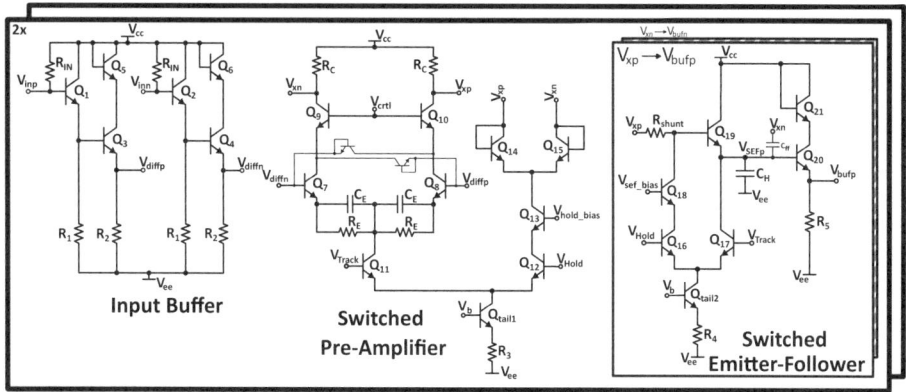

Fig. 3.5 Schematic of the high-speed THA core, implemented in an advanced SiGe technology

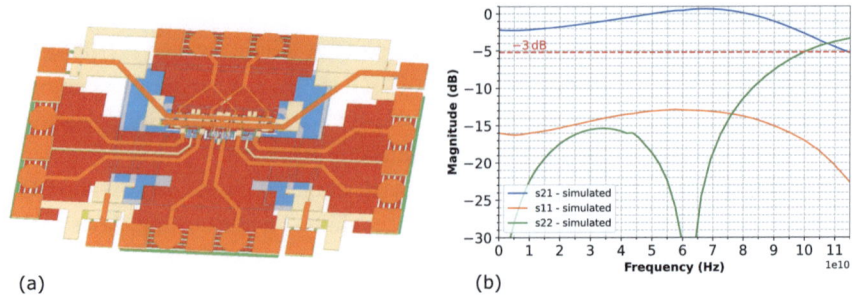

Fig. 3.6 a Layout view of the full chip in an EM simulator (ADS RFpro). **b** Post-layout simulated s-parameter results

it decreases the current of the tail-current source Q_{tail2}, which in turn reduces power consumption and helps maintain a stable hold-mode voltage during dynamic operation.

Finally, the core circuit is followed by a 50 Ω output buffer implemented as a cascode differential amplifier. With 50 Ω collector resistors and strong emitter degeneration, this stage achieves a gain of approximately 0 dB. The outputs of each channel are then connected to the chip pads via 50 Ω transmission lines.

3.2.2.3 Expected Performance

A layout for the previously described 2×-interleaved THA was created in SG13G3 technology and post-layout electromagnetic (EM) simulations were carried out. The layout of the full chip as viewed in the EM simulator can be seen in Fig. 3.6a. EM simulations include the pads and transmission lines.

Figure 3.6b shows the post-layout differential s-parameter simulations from input to one of the two outputs in track-mode. Due to the input transmission lines forming a Y-splitter, a 3 dB power split appears, which is partially compensated by the 1 dB gain in the pre-amplifier. This leads to a low-frequency gain of −2 dB, see S_{21}. The transfer function shows a 3 dB peaking at around 65 GHz and then drops to −5 dB, giving a total small-signal bandwidth of 115 GHz. Despite the formation of a Y-splitter, input reflections (S_{11}) are below −13 dB over the whole frequency band. In comparison, the output transmission lines are longer and S_{22} stays below −10 dB only up to around 82 GHz. To improve this in the future, a potential solution could be to use grounded coplanar waveguides instead of simple microstrip transmission lines. As the circuit presented here is supposed to work with a sampling frequency of 64 GS/s per channel, the most important frequency components at the output lie within the first Nyquist zone, which spans from 0 Hz to $f_s/2 = 32$ GHz and is thus not effected by the poor output matching visible at frequencies above 70 GHz.

Figure 3.7a shows the transient differential output signal of channels 1 and 2. It is clearly seen how they generate time-interleaved samples of the input signal with a sampling rate of 64 GS/s per channel. The input signal frequencies are chosen according to the requirements of coherent sampling, described in the IEEE standard 1241-2023 [23]: $f_{in} = K/N \cdot f_s$, where N is the number of samples in the record, K a prime number relative to N, and f_s the sampling rate. Specifically, in Fig. 3.7a $K = 99$, $N = 512$ and $f_s = 64\,\text{GHz}$, which result in an input frequency of 12.38 GHz. In the same way, other frequency points between 1 GHz and 110 GHz are chosen, the transient data is resampled in the middle of the hold-phases and SINAD, THD, and SNR are calculated using the sine-wave-curve-fitting method [23]. The circuit is sufficiently linear, as the THD allows for over 6 bits of effective resolution over the entire frequency band, see Fig. 3.7b. The SINAD, which takes into account noise and distortion, stays over 34 dB up to 110 GHz. The simulation results are very promising and show how a modern and fast SiGe technology allows for extremely high-speed and high-resolution samplers at moderate fabrication cost and high scalability.

3.3 Frequency-Domain Optical Arbitrary Waveform Measurement and Photonic-Electronic Analog-to-Digital Converters

In this section, several ultra-broadband signal measurement systems are presented covering functionalities of optical arbitrary waveform measurement (OAWM) and photonic-electronic analog-to-digital conversion (ADC). A common feature of all these systems is their reliance on breaking down broadband optical waveforms into multiple tributary signals, from which the original waveform is later reconstructed in the frequency domain. This is achieved either by using optical demultiplexing filters [24] or by optical down-sampling, which enables parallel signal processing in the frequency domain at relatively lower bandwidths. By leveraging the coherence of optical frequency combs, combined with dedicated signal reconstruction algorithms, these concepts allow for signal reconstruction with high-fidelity and superior bandwidth scalability.

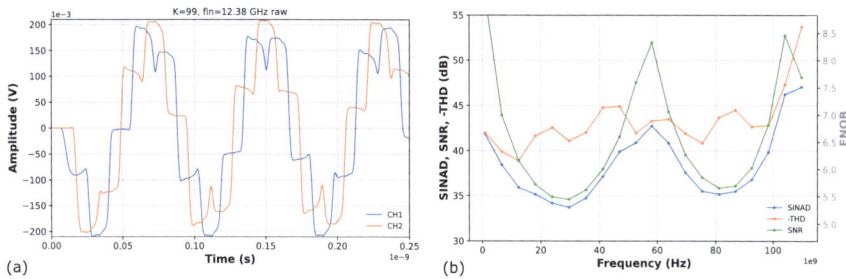

Fig. 3.7 **a** Time-domain output of channels 1 and 2. **b** Simulated SINAD, SNR, THD, and ENOB over signal frequency

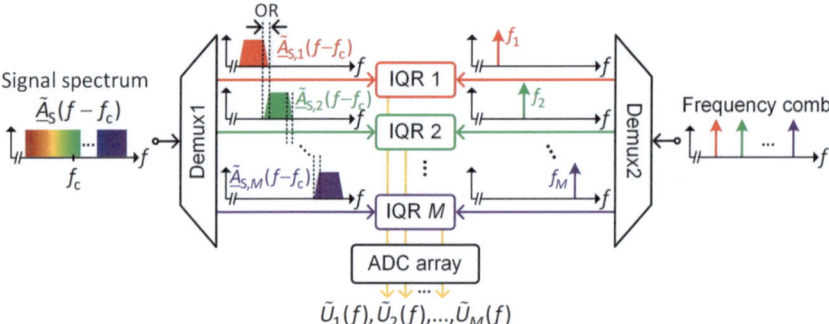

Fig. 3.8 Concept of a spectrally sliced optical arbitrary waveform measurement (OAWM) system. The incoming optical signal, described by a complex-valued frequency-domain baseband signal $\tilde{\underline{A}}_S(f)$, is decomposed by a first demultiplexer (Demux 1) into a series of signal slices $\tilde{\underline{A}}_{S,1}, \ldots, \tilde{\underline{A}}_{S,M}$, which are coherently detected by an array of in-phase and quadrature receivers (IQR 1, IQR 2,…, IQR M) using local-oscillator tones that are derived from an optical frequency comb via a second demultiplexer (Demux 2). The signal slices are digitized by an array of electronic ADCs, resulting in complex-valued signals $\tilde{\underline{U}}_1(f), \tilde{\underline{U}}_2(f), \ldots, \tilde{\underline{U}}_M(f)$. The original broadband signal is then reconstructed by DSP. For simplicity, optical amplifiers have been omitted in the illustration

3.3.1 Spectrally Sliced Optical Arbitrary Waveform Measurement (OAWM)

OAWM holds significant potential across a wide range of applications, from analyzing ultrashort events [25, 26], receiving communication signals with ultra-high symbol rates [27–29], to enabling elastic optical networking [30–32], and sliceable bandwidth-variable transponders (SBVT) in cloud radio-access networks (C-RAN) [33]. Particularly, OAWM receivers can serve as optical signal processing engines in ultra-broadband photonic-electronic ADC systems [34], where broadband incoming radio frequency (RF) signals are first converted into optical signals. Among various approaches, spectrally sliced OAWM that uses optical demultiplexing filters to subdivide the incoming broadband waveform into a series of narrowband tributaries provides several advantages, such as low crosstalk among sub-channels and straightforward scalability of the signal acquisition bandwidth that increases linearly with the number of parallelized detection channels.

3.3.1.1 Concepts

The concept of a spectrally-sliced OAWM system is illustrated in Fig. 3.8. The incoming broadband optical signal can be described as $\underline{a}_S(t) = \underline{A}_S(t)\exp(j2\pi f_c t)$, where f_c denotes the carrier frequency and $\underline{A}_S(t)$ represents the complex-valued time-domain envelope or, equivalently, the baseband representation of the signal. In the following, the signal processing steps are considered in the frequency domain, starting from the Fourier spectrum $\tilde{\underline{A}}_S(f)$ of the baseband signal. The incoming signal is first decomposed into a series of band-limited, frequency-interleaved signal slices, which are again represented via equivalent baseband spectra $\underline{\tilde{A}}_{S,\mu}(f)$, $\mu = 1, 2, \ldots, M$ as illustrated in Fig. 3.8. Note that neighboring signal slices exhibit a slight spectral overlap, indicated by the so-called overlap regions (OR) in Fig. 3.8. The redundant information contained in the OR plays a crucial role for accurately compensating unavoidable amplitude mismatches and phase drifts between adjacent detection channels in the digital signal reconstruction, see discussion below and [28] for a more detailed explanation. The spectrally sliced tributary signals are then fed to an array of in-phase and quadrature receivers (IQR 1, IQR 2,..., IQR M) for parallel coherent detection, using the spectral lines of a phase-locked optical frequency comb as multi-wavelength local oscillators (LO). The LO tones are located at frequencies f_μ, where $\mu = 1, 2, \ldots, M$, align with the center frequencies of the respective signal slices $\underline{\tilde{A}}_{S,1}(f - f_c), \underline{\tilde{A}}_{S,2}(f - f_c), \ldots, \underline{\tilde{A}}_{S,M}(f - f_c)$ and are selected and isolated from a second optical demultiplexer (Demux 2) before being routed to the corresponding IQR. An electronic ADC array is finally used to digitize the IQR outputs, producing M complex-valued baseband signals $\underline{\tilde{U}}_1(f), \underline{\tilde{U}}_2(f), \ldots, \underline{\tilde{U}}_M(f)$, from which the original waveform is reconstructed through dedicated digital signal processing (DSP) techniques.

For the signal reconstruction, it is essential to account for the fact that the measured signal slices $\underline{\tilde{U}}_1(f), \underline{\tilde{U}}_2(f), \ldots, \underline{\tilde{U}}_M(f)$ do not exactly represent the original signals. Instead, they are distorted by various components along the detection path, including the optical demultiplexing filters, optical amplifiers, which are required in a practical implementation (omitted in Fig. 3.8 for simplicity), as well as the IQRs and electronic ADCs. The cumulative impact of these linear components along each detection path can be represented by an associated transfer function $\underline{\tilde{H}}_\mu(f)$, which describes the overall system response and establishes a relationship between the original signal $\underline{\tilde{A}}_S(f)$ and the measured signal slices $\underline{\tilde{U}}_\mu(f)$,

$$\underline{\tilde{U}}_\mu(f) = \underline{\tilde{H}}_\mu(f) \underline{\tilde{A}}_S(f + f_{\text{IF},\mu}) + \underline{\tilde{N}}_\mu(f). \tag{3.3}$$

In this relation, $f_{\text{IF},\mu} = f_\mu - f_c$, $\mu = 1, 2, \ldots, M$ represents the intermediate frequencies, obtained as the difference between the associated LO-tone frequencies f_μ and the carrier frequency f_c of the incoming waveform [28, 34]. These frequency shifts result from down-conversion of the respective tributary signals with corresponding LO tones at frequencies f_μ. Additionally, the measured signal slices $\underline{\tilde{U}}_\mu(f)$ are assumed to be impaired by statistically independent white Gaussian noise (AWGN) $\underline{\tilde{N}}_\mu(f)$ for each slice μ, see [28] for details. In the initial stage of signal reconstruction, the detected baseband signal slices are shifted to their respective intermediate center frequencies $f_{\text{IF},\mu}$,

$$\underline{\tilde{U}}_\mu(f - f_{\text{IF},\mu}) = \underline{\tilde{H}}_\mu(f - f_{\text{IF},\mu}) \underline{\tilde{A}}_S(f) + \underline{\tilde{N}}_\mu(f - f_{\text{IF},\mu}). \tag{3.4}$$

For a system with M slices, the equations for the various μ can be re-written by using $(M, 1)$ column matrices,

$$\underbrace{\begin{bmatrix} \tilde{U}_1(f-f_{\text{IF},1}) \\ \tilde{U}_2(f-f_{\text{IF},2}) \\ \vdots \\ \tilde{U}_M(f-f_{\text{IF},M}) \end{bmatrix}}_{\tilde{\underline{U}}(f)} = \underbrace{\begin{bmatrix} \tilde{H}_1(f-f_{\text{IF},1}) \\ \tilde{H}_2(f-f_{\text{IF},2}) \\ \vdots \\ \tilde{H}_M(f-f_{\text{IF},M}) \end{bmatrix}}_{\tilde{\underline{H}}(f)} \tilde{A}_S(f) + \underbrace{\begin{bmatrix} \tilde{N}_1(f-f_{\text{IF},1}) \\ \tilde{N}_2(f-f_{\text{IF},2}) \\ \vdots \\ \tilde{N}_M(f-f_{\text{IF},M}) \end{bmatrix}}_{\tilde{\underline{N}}(f)}. \qquad (3.5)$$

If the transfer functions $\tilde{\underline{H}}(f)$ is known, the original signal $\tilde{A}_S(f)$ can be estimated by using the maximum-ratio combining (MRC) algorithm,

$$\tilde{A}_S^{(\text{est})}(f) = \left(\tilde{\underline{H}}^\dagger(f)\tilde{\underline{H}}(f)\right)^{-1}\tilde{\underline{H}}^\dagger(f)\tilde{\underline{U}}(f), \qquad (3.6)$$

where $\left(\tilde{\underline{H}}^\dagger(f)\tilde{\underline{H}}(f)\right)^{-1}\tilde{\underline{H}}^\dagger(f)$ is the pseudo-inverse of the column matrix $\tilde{\underline{H}}(f)$, see [28] for details. Without noise, the broadband input signal $\tilde{A}_S^{(\text{est})}(f) = \tilde{A}_S(f)$ is reconstructed exactly this way. Otherwise, $\tilde{A}_S^{(\text{est})}(f)$ provides a least-square error estimation of the optical baseband signal $\tilde{A}_S(f)$.

In a practical implementation, the key challenge lies in accurately measuring the complex-valued transfer functions $\tilde{\underline{H}}(f)$, which are impacted by amplitude and phase drifts of the comb tones and random phase variations in fibers. This leads to slow variations of transfer functions over time. Since these variations are rather slow compared to the typical μs-scale recording time, the transfer functions can be considered constant within a single recording but may vary between subsequent ones. This can be modeled by splitting each element of the transfer function column matrix $\tilde{\underline{H}}(f)$ into a time-invariant frequency-dependent part, corresponding to a column matrix $\tilde{\underline{H}}_f(f)$, and into a time-variant part represented by the column matrix $\tilde{\underline{H}}_\tau$ that does not depend on frequency,

$$\tilde{\underline{H}}(f) = \tilde{\underline{H}}_f(f) \odot \tilde{\underline{H}}_\tau, \qquad (3.7)$$

where the Hadamard product $\odot$ represents the element-wise multiplication of the two column matrices $\tilde{\underline{H}}_f(f)$ and $\tilde{\underline{H}}_\tau$. The time-invariant frequency-dependent part $\tilde{\underline{H}}_f(f)$ is obtained from a one-time calibration measurement by feeding the system with a known optical reference waveform, derived from an ultra-stable femtosecond laser with well-known pulse shape, see [28] for details. The time-variant part $\tilde{\underline{H}}_\tau$, however, must be estimated for each recording. This can be achieved by comparing the redundantly measured complex-valued spectral components in the OR between neighboring slices, see Fig. 3.8. A more detailed explanation of this procedure can be found in [28].

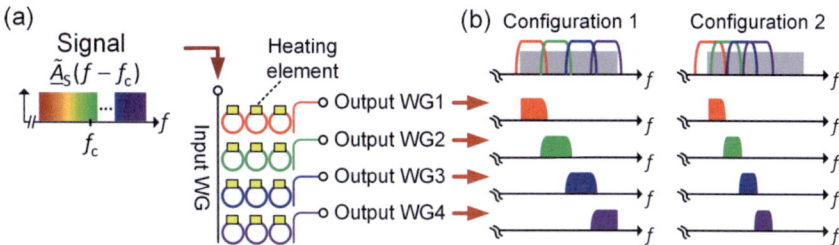

Fig. 3.9 Gapless slicing with adjustable slice width using a bank of cascaded CROW filters. **a** The signal $\tilde{A}_S(f - f_c)$ is fed into the input bus waveguide (Input WG), and spectrally sliced by four stages of CROW filters. Each stage consists of 3 rings with heating elements for thermal tuning, and is coupled to its own output waveguides (Output WG1...Output WG4). **b** The spectral widths of the individual signal slices can be flexibly adjusted by appropriate tuning of the first CROW filter (red) with respect to the edge of the signal spectrum (gray) and by adjusting the center frequencies of all subsequent CROW filters according to the desired slice widths. Gapless slicing is achieved by spectral overlap of the various filter passbands, such that spectral components that are not extracted by one CROW filter may be captured by the next one

3.3.1.2 Demonstrations

Clearly, spectrally sliced OAWM systems can greatly benefit from photonic integration, in particular when exploiting advanced integration platforms that offer tunability of the underlying passive photonic circuits. This was demonstrated by implementing a silicon photonic OAWM receiver that exploits a bank of highly flexible frequency-tunable coupled-resonator optical waveguide (CROW) filters for spectral slicing of the incoming signal [28]. The CROW filters offer wideband tunability along with flat-top transmission and steep roll-off, and are particularly well suited for gapless spectral slicing with adjustable slice width. Figure 3.9a shows a sketch of a four-channel CROW filter bank, where each CROW filter stage consists of three ring resonators that can be thermally tuned via heating elements. The four groups of rings are coupled to a common input bus waveguide (Input WG) while having their own output waveguides (Output WG1...Output WG4). The incoming signal $\tilde{A}_S(f - f_c)$ is fed into the input waveguide and reaches the four CROW filters successively. This configuration allows to flexibly adjust the spectral widths of the individual signal slices by appropriate tuning of the first CROW filter with respect to the edge of the signal and by adjusting the center frequencies of all subsequent filters according to the desired slice widths. In other words, dynamic reconfiguration of the spectral slice widths can be achieved by only shifting the center frequencies of the various filters, without adjusting the bandwidth of the filters themselves. Two configurations that achieve different slice widths are illustrated in Fig. 3.9b.

For an experimental demonstration, the CROW filters were fabricated on the silicon photonic platform and subsequently packaged with both electrical and optical interfaces [28]. The four LO tones were derived from an optical frequency comb featuring a free spectral range (FSR) of 35 GHz, generated by electro-optic modulation. For simplicity, the second

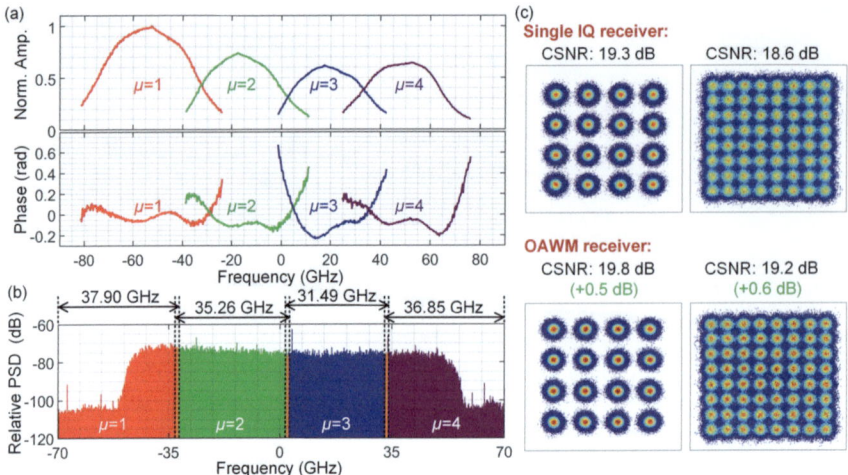

Fig. 3.10 Experimental results of a spectrally sliced OAWM system using silicon photonic CROW filters. **a** Amplitudes and phases of the measured frequency-dependent transfer functions. The amplitudes have been normalized to the maximum value obtained for the first slice. **b** Reconstructed spectrum of a 100 GBd 64QAM signal, obtained through merging the signal slices by maximum-ratio combining (MRC). The ORs have widths of only 500 MHz (small orange-colored regions), which is sufficient for signal reconstruction. The bandwidths of various slices are slightly different due to the non-uniform transfer characteristics of the CROW filters. **c** Comparison of 100 GBd optical data signals obtained by a single-slice single-polarization intradyne coherent receiver (Single Rx) and by OAWM-based detection (OAWM). The constellation signal-to-noise ratio (CSNR) for all 100 GBd QAM signals is slightly better than that of the conventional single-slice receiver, while the requirements with respect to the electronic acquisition bandwidths of the individual ADC are greatly reduced. Figure adapted from [28]

optical demultiplexer (Demux 2) and the IQR array still relied on off-the-shelf fiber-optic components. The frequency-dependent part of the transfer functions $\tilde{\underline{\mathbf{H}}}_f(f)$ of the various detection paths were measured using the aforementioned one-time calibration method and are displayed in Fig. 3.10a. To test the performance of our OAWM receiver, 100 GBd optical communication signals with different modulation formats were used as test signals. An exemplary reconstructed signal obtained from four tributary signals is shown in Fig. 3.10b. In this case, the spectral widths of the four signal slices amount to 37.90 GHz, 32.26 GHz, 31.49 GHz, and 36.85 GHz, including the spectral overlap region to one (Slices 1 and 4) or to both (Slices 2 and 3) sides of the respective slice. This leads to an overall optical detection bandwidth of 140 GHz for the demonstrated OAWM system. To estimate the quality of the reconstructed signal, the results were compared to those obtained from a conventional high-speed coherent receiver. Figure 3.10c shows a comparison of the corresponding 16QAM and 64QAM constellation diagrams along with the associated data-aided constellation signal-to-noise ratio (CSNR), obtained by using identical DSP algorithms in both cases. The results show that the OAWM receiver offers better signal quality for both 16QAM and 64QAM modulation formats, which can be attributed to less electronic noise being captured in each individual slice due to the four-fold reduced electronic bandwidth needed for the individual ADCs. Note that the scheme can be easily scaled to higher bandwidths by increasing the number of spectral slices. The use of an integrated silicon photonic spectral slicer also represents an important step towards a fully integrated on-chip OAWM system that can be used in practical application scenarios outside well-controlled laboratory environment.

3.3.2 Non-Sliced Optical Arbitrary-Waveform Measurement (OAWM)

As shown in the previous subsection, spectrally sliced OAWM techniques can overcome the bandwidth limitation of conventional coherent detection schemes. However, chip-level integration of high quality optical slicing filters is challenging and typically requires complex tuning procedures to achieve the desired performance. To overcome these challenges, a concept referred to as non-sliced OAWM has been proposed and demonstrated [35], as will be explained in the following. The non-sliced scheme avoids optical slicing filters and instead relies on simple wavelength-agnostic power splitters and optical delay lines–components which are well-suited for chip-level integration using high index-contrast platforms such as silicon photonics [29].

3.3.2.1 Concepts

Figure 3.11a shows a conceptual setup for non-sliced OAWM. The incoming optical signal $\underline{a}_S(t) = \underline{A}_S(t)\exp(j2\pi f_c t)$ is again described by its optical carrier frequency f_c and by the associated complex-valued time-domain envelope $\underline{A}_S(t)$ and its Fourier spectrum $\underline{\tilde{A}}_S(f)$. At the input of the OAWM system, the incoming signal is split into N copies by a wavelength-agnostic splitter. The various signal copies are then routed to an array of in-phase and quadrature receivers (IQRs), which are labeled by subscripts $\nu = 1, ..., N$. These IQRs are also fed with copies of an optical frequency comb that undergo dedicated time delays τ_ν and that serve as local-oscillator (LO) signals. These LO signals comprise M comb tones at optical frequencies f_μ, $\mu = 1, ..., M$, separated by a free spectral range (FSR) f_{FSR}, see the right-hand side of Fig. 3.11a. Each IQR-ν generates an in-phase signal $I_\nu(t)$ and a quadrature signal $Q_\nu(t)$ at its outputs. These output signals are acquired by corresponding electrical ADCs and can be combined into a single complex-valued baseband signal $\underline{U}_\nu(t) = I_\nu(t) + jQ_\nu(t)$, assuming that the frequency responses of the I- and Q-channel are identical and that both quadratures have an ideal 90° phase relationship, see supplementary document of [35] for a detailed discussion. Subsequently, digital signal processing (DSP) is used to reconstruct the spectrum $\underline{\tilde{A}}_S(f)$ of the complex-valued envelope $\underline{A}_S(t)$ of the optical input signal from the N measured complex-valued baseband signals $\underline{U}_\nu(t)$, Fig. 3.11b.

Because the optical input signal with spectrum $\underline{\tilde{A}}_S(f - f_c)$ is simultaneously down-converted by all M LO comb tones, each measured baseband spectrum $\underline{\tilde{U}}_\nu(f)$ contains a superposition of M frequency-shifted copies $\underline{\tilde{A}}_S(f - f_c + f_\mu)$, $\mu = 1, ..., M$ of the input spectrum $\underline{\tilde{A}}_S(f - f_c)$, where the frequency shift is dictated by the respective LO tone frequency f_μ and where the detection bandwidth B is limited by the bandwidth of the IQRs and the ADCs. Note that if the detection bandwidth B exceeds half of the FSR of the LO comb, $B > f_{FSR}/2$, then spectral components of the input signal located in the overlap region (OR) appear twice in the measured baseband spectrum $\underline{\tilde{U}}_\nu(f)$–once at positive and once at negative frequencies, Fig. 3.11b. This redundancy is exploited for estimating unknown

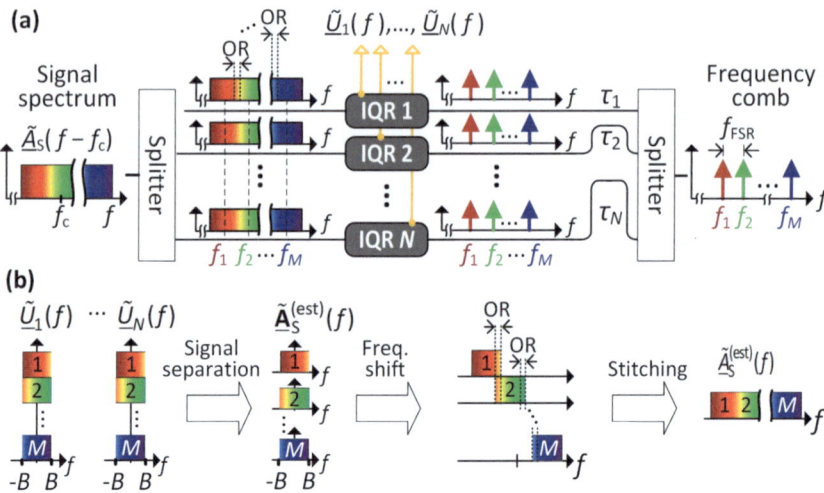

Fig. 3.11 Concept of non-sliced OAWM. **a** A broadband optical input signal $\underline{A}_S(t)\exp(j2\pi f_c t)$ with spectrum $\underline{\tilde{A}}_S(f - f_c)$ and carrier frequency f_c is split into N copies and fed to an array of in-phase and quadrature receivers (IQRs). The IQRs are also fed with delayed copies (delays $\tau_1, ..., \tau_N$) of a frequency comb that is used as a multi-wavelength local oscillator (LO). The simultaneous downconversion with all M LO comb tones leads to a superposition of M spectral components in the complex-valued baseband signals $\underline{U}_\nu(t) = I_\nu(t) + jQ_\nu(t)$, $\nu = 1, ..., N$ that are extracted from the N IQRs. Spectral components within the overlap regions (OR) are downconverted by both adjacent LO comb tones if the detection bandwidth B exceeds half the FSR of the LO, $B > f_{FSR}/2$. **b** Starting from the spectra $\underline{\tilde{U}}_\nu(f)$ of the captured baseband signals $\underline{U}_\nu(t)$, the signal spectrum $\underline{\tilde{A}}_S(f)$ is reconstructed in three steps: First, the superposition of the various spectral components (labeled 1, 2,..., M) in the measured baseband spectra $\underline{\tilde{U}}(f) = [\underline{\tilde{U}}_1(f), ..., \underline{\tilde{U}}_M(f)]^T$ is undone by multiplying the inverse transfer matrix $\underline{\tilde{H}}^+(f)$ with the measured baseband spectra $\underline{\tilde{U}}(f)$, see Eq. 3.10. The resulting signal vector $\underline{\tilde{A}}_S^{(est)}(f)$ comprises M spectral slices, which in a second step are frequency shifted according to the FSR of the LO comb. Assuming $B > f_{FSR}/2$, the resulting spectral slices are overlapping within the overlap regions (OR). In the third step, all spectral slices are combined according to Eq. 3.11, resulting in the reconstructed signal spectrum $\underline{\tilde{A}}_S^{(est)}(f)$. Note that the signal reconstruction is sensitive to optical phase drifts between the various detection paths. These phase drifts are estimated and compensated by exploiting redundantly measured signal components in the overlap regions, see [35] for details

system parameters such as phase drifts in the optical fibers leading to the various IQRs, see [35] for a more detailed discussion.

Similarly to the model used for spectrally sliced OAWM, the frequency response of the ADCs, the IQRs, the optical transfer characteristics of the various detection paths, as well as the complex-valued amplitudes of the individual LO comb tones can be combined into equivalent transfer functions $\underline{\tilde{H}}_{\nu\mu}(f)$, relating the frequency-shifted signal portions $\underline{\tilde{A}}_S(f - f_c + f_\mu)$ to the measured baseband spectra $\underline{\tilde{U}}_\nu(f)$ via a matrix vector equation [35],

$$\underbrace{\begin{pmatrix} \tilde{U}_1(f) \\ \vdots \\ \tilde{U}_N(f) \end{pmatrix}}_{\underline{\tilde{U}}(f)} = \underbrace{\begin{pmatrix} \tilde{H}_{11}(f) & \cdots & \tilde{H}_{1M}(f) \\ \vdots & \ddots & \vdots \\ \tilde{H}_{N1}(f) & \cdots & \tilde{H}_{NM}(f) \end{pmatrix}}_{\underline{\tilde{H}}(f)} \underbrace{\begin{pmatrix} \tilde{A}_S(f - f_c + f_1) \\ \vdots \\ \tilde{A}_S(f - f_c + f_M) \end{pmatrix}}_{\underline{\tilde{A}}_S(f)}. \quad (3.8)$$

The signal reconstruction now essentially relies on the inversion of the above relation, which is possible if there are at least as many IQRs N as there are superimposed spectral portions M, i.e., Eq. 3.8 must not be underdetermined. Additionally, the transfer matrix $\underline{\tilde{H}}(f)$ must be well conditioned, which can be achieved by choosing approximately equidistant delays τ_ν in the LO path with respect to the repetition period $T_{LO} = 1/f_{FSR}$ of the LO comb [35],

$$\tau_\nu \approx \frac{(\nu - 1)T_{LO}}{N}, \quad \nu = 1, 2, \ldots N. \quad (3.9)$$

To calculate a least-squares estimate $\underline{\tilde{A}}_S^{(est)}(f) = [\tilde{A}_{S,1}^{(est)}(f), \ldots, \tilde{A}_{S,M}^{(est)}(f)]^T$ of the signal vector $\underline{\tilde{A}}_S(f)$ within the receiver bandwidth B we can then use the pseudo-inverse $\underline{\tilde{H}}^+ = (\underline{\tilde{H}}^\dagger \underline{\tilde{H}})^{-1} \underline{\tilde{H}}^\dagger$, see [35] for details,

$$\underline{\tilde{A}}_S^{(est)}(f) = \underline{\tilde{H}}^+(f) \underline{\tilde{U}}(f) \quad \text{for } |f| < B. \quad (3.10)$$

Note that for the special case where the number of IQRs equals the number of LO comb tones, $N = M$, the pseudo inverse in the above relation simplifies to the normal matrix inverse, i.e., $\underline{\tilde{A}}_S^{(est)}(f) = \underline{\tilde{H}}^{-1}(f) \underline{\tilde{U}}(f)$.

Once the estimated signal vector $\underline{\tilde{A}}_S^{(est)}(f)$ has been obtained according to Eq. 3.10, its components $\tilde{A}_{S,\mu}^{(est)}(f)$ are frequency-shifted back by $f_\mu - f_c$, see Fig. 3.11b. Subsequently, weighting functions $W_\mu(f)$ fulfilling the relation $\sum_\mu W_\mu(f) = 1$ are used before summing all spectral components,

$$\underline{\tilde{A}}_S^{(\text{est})}(f) = \sum_\mu W_\mu(f) \underline{\tilde{A}}_{S,\mu}^{(\text{est})}(f - f_\mu + f_c). \tag{3.11}$$

In the last step, an inverse Fourier transform is applied to obtain the complex-valued time-domain amplitude $\underline{\tilde{A}}_S^{(\text{est})}(t)$. Further details of the signal reconstruction techniques for non-sliced OAWM can be found in [35].

Note that the signal reconstruction relies on the precise knowledge of the transfer functions $\underline{\tilde{H}}_{\nu\mu}(f)$, which include the frequency responses of all system components in the signal path, e.g., optical waveguides, photodetectors, and ADCs. The various frequency dependent transfer functions $\underline{\tilde{H}}_{\nu\mu}(f)$ are determined once in a dedicated calibration measurement [29, 35]. However, as explained above for the case of sliced OAWM, slowly time-variant optical phase drifts between the various detection paths lead to frequency-independent phase offsets of the various transfer functions and require additional phase estimation and compensation techniques [35]. This is achieved by again exploiting redundant information that can be found in the complex-valued baseband signals $\underline{\tilde{U}}_\nu(f)$ if the bandwidth B of the various detection channels exceeds half of the LO comb's FSR, $B > f_{\text{FSR}}/2$ [35]. More specifically, after recovering the signal vector components $\underline{\tilde{A}}_{S,\mu}^{(\text{est})}(f)$ and applying the required frequency shift $f_\mu - f_c$, the resulting spectral slices exhibit spectral overlap regions (OR) as illustrated in Fig. 3.11b, containing signal components that originate from the same spectral portion of the incoming signal. Unknown system parameters such as phase offsets can hence be found by minimizing amplitude and phase errors between the overlapping spectral components, see [35] and supplementary document thereof for a detailed discussion.

It is also noteworthy that in the case of an undispersed comb with Fourier-transform-limited pulses, the architecture described in this section reduces to coherent sampling of a broadband signal with pulse trains to which different delays have been applied [36, 37], not so different from the opto-electronic sampling architectures presented in Sect. 3.2 at an abstract level. This consideration also creates a basis for the analogies drawn in Sect. 3.4 in regard to this architecture.

3.3.2.2 Demonstrations

The viability and bandwidth scalability of non-sliced OAWM was demonstrated by conducting a proof-of-concept experiment based on four high-bandwidth fiber-optic IQRs and a frequency comb source with an FSR in excess of 150 GHz [35, 38]. This configuration allowed for an aggregate acquisition bandwidth of 610 GHz, which was used to receive an optical test signal comprising seven quadrature amplitude modulated (QAM) signals. Figure 3.12 shows the reconstructed spectrum along with the constellation diagrams associated with the seven QAM signals. The constellation signal-to-noise ratio (CSNR) lays between 18 dB and 21 dB for all channels and can potentially be further improved, e.g., by using an optical frequency comb with a higher optical carrier-to-noise ratio (OCNR) and by employing photodetectors with a sufficient bandwidth for all detection channels [35].

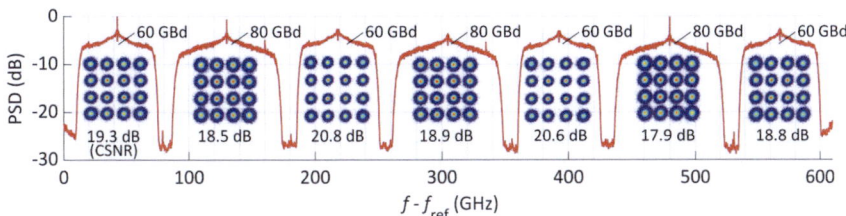

Fig. 3.12 Experimental demonstration of the acquisition of a 610 GHz test signal using non-sliced OAWM [35]. The acquired test signal comprises seven QAM signals at symbol rates of 60 GBd and 80 GBd. All QAM signals are demodulated, and constellation signal-to-noise ratio (CSNR) levels between 18 dB and 21 dB are achieved. Note that the horizontal axis indicates the offset of the optical frequency f from a reference frequency $f_{ref} = 192.52$ THz, which was chosen to correspond to the lower frequency edge of the first spectral slice

A second demonstration exploits the amenability of non-sliced OAWM schemes to photonic integration and relies a compact (1.5×1.0 mm^2) silicon photonic integrated circuit (PIC) [29, 39]. The PIC comprises cascaded 2×2 multimode interference (MMI) couplers for splitting the signal and LO into four paths, as well as four optical delay lines for the LO (τ_1, τ_2, τ_3, and τ_4) and four corresponding IQRs. Each IQR consists of a 2×4 MMI coupler that serves as a 90° optical hybrid and two balanced germanium-based photodetectors, see Fig. 3.13a. The PIC was tested by contacting the photodetectors via a high-frequency probe and by recording the generated signals with high-bandwidth oscilloscopes, see [29] for details. All contacted photodetectors offered a 3 dB bandwidth of approximately 15 GHz and a 6 dB bandwidth of approximately 22 GHz [29]. The performance of the four-channel OAWM PIC was demonstrated by acquiring a 160 GHz optical test signal that was synthesized by combining an 80 GBd and a 60 GBd 64QAM signal. Figure 3.13b shows the reconstructed signal spectrum as well as the associated constellation diagrams. As a reference, the same test signal was measured using a single-channel IQ receiver that relies on two 100 GHz high-end balanced photodetectors, Fig. 3.13b. It was found that the OAWM system offers a better signal quality compared to the single-channel receiver, because it relies on ADCs with lower bandwidth that offer a higher ENOB.

3.3.3 Photonic-Electronic Analog-to-Digital Converter (PE-ADC)

Combining OAWM systems with high-speed electro-optic modulators finally paves a path towards ultra-broadband ADCs. While BiCMOS or CMOS-based ADCs combine compactness and robustness with the inherent scalability of the underlying technology platforms, the acquisition bandwidths of these devices are limited to typically less than 60 GHz with sampling rates of less than 200 GSa/s [40]. The bandwidth of individual ADC can be increased by electronic time-domain or frequency-domain multiplexing, leading to overall

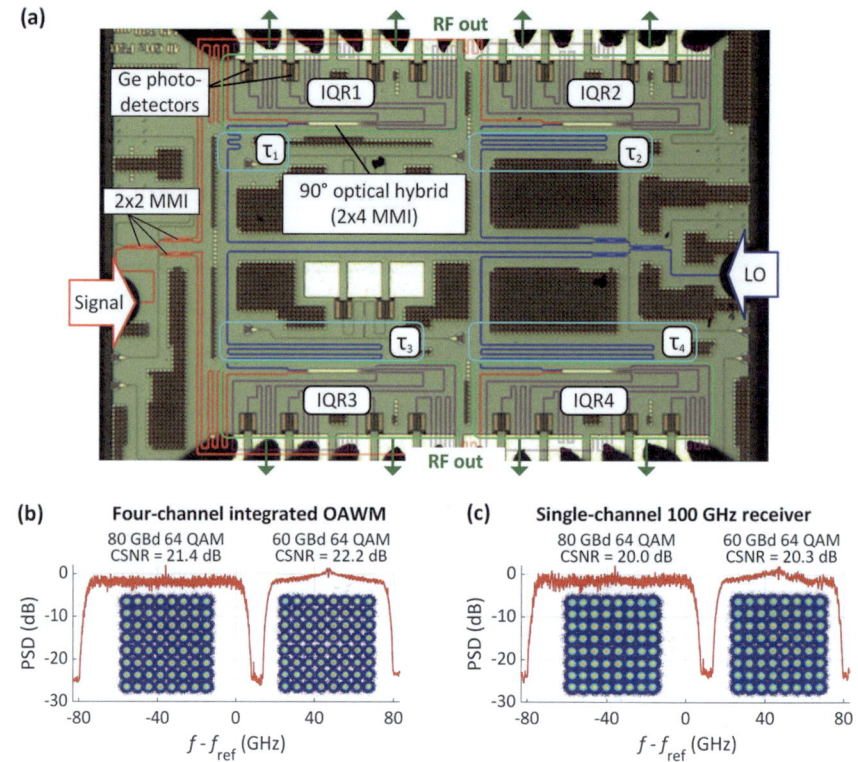

Fig. 3.13 Demonstration of non-sliced OAWM based on a compact silicon photonic PIC. **a** Microscope image of the OAWM PIC comprising 2×2 and 2×4 multi-mode interference (MMI) couplers as power splitters and 90° optical hybrids, respectively, optical delay lines ($\tau_1 \ldots \tau_4$), as well as IQ receivers (IQR1 ... IQR4) based on germanium photodetectors. **b** Test signal comprising two 64QAM data channels acquired using the four-channel OAWM PIC shown in panel (a). **c** Test signal acquired using a single-channel coherent receiver. (Panels (a) and (b) are adapted from [29] under CC-BY 4.0 license)

real-time acquisition bandwidths of up to 110 GHz with an effective number of bits (ENOB) of approximately 5 [41]. However, the bandwidth-scalability of electronic demultiplexing schemes is limited by the analog multiplexer circuits and by the jitter of the associated RF oscillators [2]. Photonic-electronic ADCs may overcome these limitations [2, 4, 34, 42, 43]. In this section, we present an ultra-broadband photonic-electronic ADC based on the spectrally sliced OAWM technique described in Sect. 3.3.1 above.

3.3.3.1 Concepts

Our scheme relies on a high-speed electro-optic (EO) Mach-Zehnder modulator (MZM) that first converts the broadband analog electrical waveform to an optical signal, which is then detected via spectrally sliced OAWM as described in the Sect. 3.3.1 above. Precise reconstruction of the electrical waveform does not only require careful calibration of the OAWM system, but also a measurement of the transfer function related to the MZM.

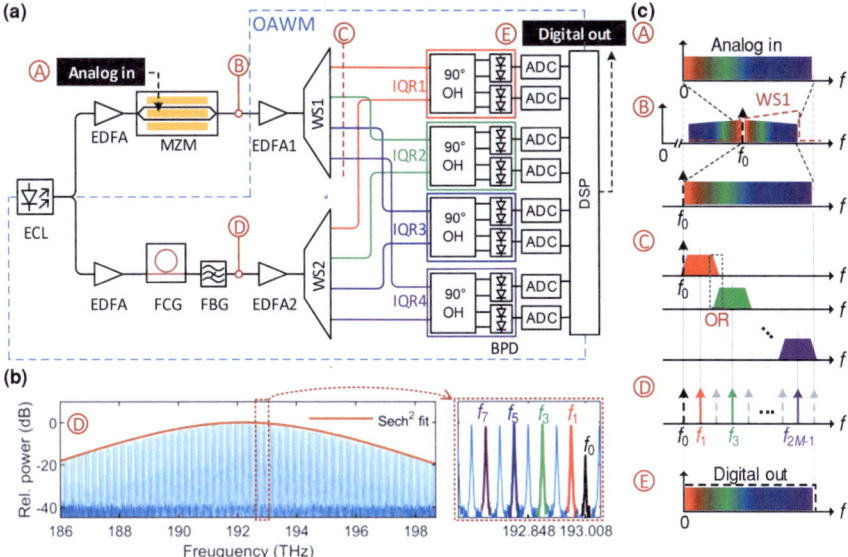

Fig. 3.14 a Concept and simplified sketch of the experimental setup of a photonic-electronic ADC that exploits spectrally sliced OAWM as a broadband signal-acquisition technique. An external-cavity laser (ECL) is used to generate the carrier for optical modulation and Kerr-comb generation. Electro-optic conversion of the analog electrical input signal (Analog in) is accomplished by a thin-film lithium-niobate (TFLN) Mach-Zehnder modulator (MZM). Programmable optical filters (WS) are used for slicing of the resulting optical signal and the corresponding LO comb, which is derived from an Si_3N_4-based Kerr comb generator (FCG, $f_{FSR} = 40.025$ GHz) [44–46]. The four spectral slices are detected by an array of optical IQ receivers (IQR1,..., IQR4), which comprise 90° optical hybrids (OH) and balanced photodetectors (BPD). The overall eight BPD are read out by eight synchronized ADC channels. A fiber Bragg grating (FBG) is used to suppress the remaining pump tone after the FCG, and erbium-doped fiber amplifiers (EDFA) help to overcome the optical losses in the setup. **b** Spectrum of the single-soliton Kerr comb along with a zoom-in of the selected LO tones. Note that the lower sideband of the signal spectrum was detected in the experiment due to the slightly better optical carrier-to-noise ratio (OCNR) of the comb tones, while the spectral scheme in Subfigure (c) illustrates the case of upper sideband detection for easier understanding. **c** Illustration of the spectrally sliced photonic-electronic ADC in the spectral domain: Ⓐ Broadband analogue electrical input signal. Ⓑ Resulting optical signal, generated by modulating the broadband analogue electrical input signal onto an optical carrier, prior to slicing into M tributaries by WS1. Ⓒ Spectrally sliced signal tributaries with small overlap regions (OR) between neighboring slices, generated at the output of WS1. Ⓓ Spectrum of the frequency comb prior to slicing by WS2. The selected tones are color-coded according to the corresponding signal slices. Ⓔ Reconstructed spectrum of the analogue electrical input signal, obtained by merging the *M* spectral slices in the digital domain. (Figure adapted from [34] under CC-BY 4.0 license)

The concept of the experimental setup of the photonic-electronic ADC is illustrated in Fig. 3.14a. For conversion of the analog electrical input signal (Analog in), a high-speed thin-film lithium-niobate (TFLN) MZM biased at the zero-transmission point is exploited. The TFLN MZM can handle high optical powers and can reach 3 dB bandwidths beyond 100 GHz while still showing a reasonable modulation response well beyond 300 GHz [47, 48]. To accurately measure and reconstruct broadband electrical input signals, the electro-optic (EO) response of the MZM must be compensated for. To this end, sinusoidal signals with known RF powers and frequencies up to 320 GHz are fed to the MZM and the EO response can be obtained by dividing the optical sideband power by the known RF input power, see Supplementary Information (SI) of reference [34] for a detailed discussion. For simplicity, we use discrete fiber-optic components such as a programmable wave shaper (WS) for spectral slicing to implement the OAWM subsystem. The multi-wavelength LO for parallel coherent detection is provided by a Kerr-soliton microcomb with an FSR of $f_{FSR} = 40.025$ GHz, derived from an Si_3N_4-based chip-scale Kerr soliton microcomb [44–46]. Such combs can easily provide hundreds of phase-locked comb lines while featuring ultra-low timing jitter. The spectrum of the generated comb and a zoom-in of the selected LO tones used for the experiment are shown in Fig. 3.14b. The frequency-domain signal stitching scheme is illustrated in Fig. 3.14c, in which electrical and optical signal spectra that occur in the architecture concept shown in Fig. 3.14a are color-coded to illustrate how they correspond to each other. Note that the same external-cavity laser (ECL) provides the carrier for the MZM and the pump for the Si_3N_4 comb generator such that phase noise of the signal and of the LO essentially cancel. Since the spectral width of each optical signal slice corresponds to twice the native FSR f_{FSR} of the comb, only every second tone positioned in the center of a slice at frequency $f_1, f_3, \ldots, f_{2M-1}$ is used as an LO tone for coherent reception. This ensures gapless signal acquisition from DC to an upper frequency of $2 f_{FSR} \times M$. For the implementation with $M = 4$ spectral slices along with a comb FSR of $f_{FSR} = 40.025$ GHz, an acquisition bandwidth of 320 GHz in the first Nyquist band can be achieved, corresponding to an effective sampling rate of at least 640 GSa/s.

3.3.3.2 Demonstrations

The viability of the photonic-electronic ADC concept was demonstrated in two sets of experiments. In a first set of experiments, sinusoidal electrical test signals with frequencies of 56 GHz, 280.8 GHz, and 307.8 GHz were generated using different signal generators, and then fed to the analogue input of the photonic-electronic ADC system while the associated digital signal was recorded, see Fig. 3.15a. To investigate the quality of the measured signals, a sinc-interpolation of the sampling points (green circles/blue dots) was performed and a sinusoidal model function (red line) was fitted to the measured sampling points to approximate the associated continuous-time waveforms. The deviations of the interpolated points from the sinusoidal fits are minute, underlining the quality of the reconstructed waveforms. In all of the cases illustrated in Fig. 3.15a, all four OAWM slices were reconstructed and arti-

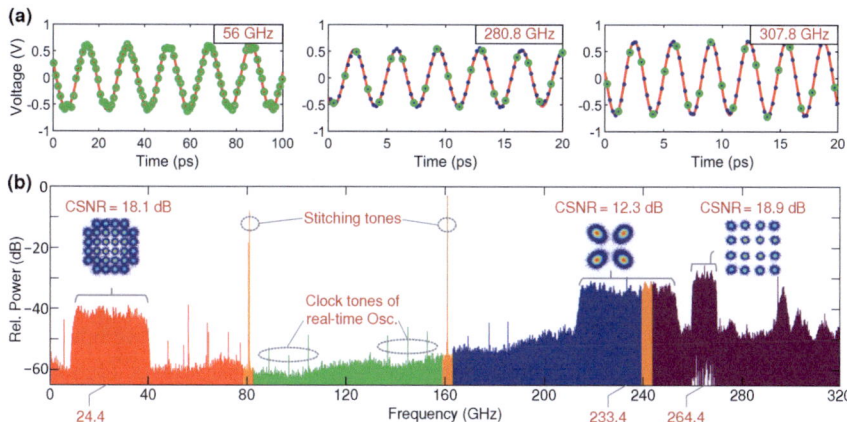

Fig. 3.15 Experimental results of a spectrally sliced photonic-electronic ADC. **a** Digitized sinusoidal test signals at frequencies of 56 GHz, 280.8 GHz, and 307.8 GHz. Green circles: Sampling points. Red curve: Least-squares fit of a sinusoidal model function. Blue dots (only in two graphs): Sinc-interpolated sampling points. The deviations of the interpolated points from the sinusoidal fits are minute, underlining the quality of the reconstructed waveforms. **b** Digitized broadband electronic signal. The electronic signal comprises three QAM data signals, centered at carrier frequencies 24.4 GHz (30 GBd 32QAM), 233.4 GHz (40 GBd QPSK), and 264.4 GHz (10 GBd 16QAM). From the digital representation indicated by horizontal braces, three constellation diagrams can be extracted. In the empty region, clock tones from the real-time oscilloscope are visible. In the middle of the empty overlap regions, stitching tones (orange) help to adjust the phases of neighboring slices. (Figure adapted from [34] under CC-BY 4.0 license)

ficial stitching tones in the optical domain were added to facilitate spectral stitching in case the spectral overlap region between neighboring slices does not contain any components of the use signal. Note also that our system operates within the first Nyquist band–unlike other previously demonstrated photonic ADC systems, which relied on down-sampling of single-frequency tones in higher-order Nyquist bands [2, 42].

Offering the full bandwidth of 320 GHz in the first Nyquist band is key to the acquisition of broadband analogue test signals. In a second experiment, the photonic-electronic ADC was used to acquire a broadband analogue data signal, consisting of a 30 GBd 32QAM waveform centered at 24.4 GHz, a 40 GBd QPSK waveform centered at 233.4 GHz, and a 10 GBd 16QAM waveform centered at 264.4 GHz. Figure 3.15b shows the spectrum of the reconstructed signal along with the constellation diagrams of the data signals extracted by applying state-of-the-art DSP techniques to the digitized signal. The constellation signal-to-noise ratio (CSNR) of the 30 GBd, 40 GBd, and 10 GBd data signals amount to 18.1 dB, 12.3 dB, and 18.9 dB, respectively. By increasing the number of slices and by exploiting highly stable optical frequency combs with ultra-low phase noise [49], our concept may open a path to acquisition bandwidths in the order of 1 THz using a fully integrated photonic-electronic system [28].

3.4 Optical Deserializers with Built-In Equalization

As an outlook for the systems explored above, we are giving a brief introduction to an architecture expanding the concept of the filter-less OAWM described in Sect. 3.3.2 to a configuration aimed specifically at the deserialization and equalization of a broadband signal generated by a co-designed transmitter. A more detailed analysis including optical power budget, fabrication tolerance and comb source requirement analyses can be found in [6, 7].

Figure 3.16 gives an overview of the system architecture, with Fig. 3.16a depicting the transmitter (Tx) and Fig. 3.16b depicting the receiver (Rx). The architecture is in particular adapted such that the broadband signal and the reference comb used to analyze it can be sent down through the same fiber between the Tx and Rx. This greatly enhances phase stability between these two optical signals, since phase perturbations occuring in the fiber are applied to both. To make this possible, the broadband signal is not applied to a solitary carrier, but rather to an interleaved train of pulses generated from the same comb. This is depicted in the lower branch of the Tx architecture, in which four pulses are first exemplarily interleaved, enabling a deserialization by a factor $N = 4$ downstream in the Rx. This is followed by phase modulation, that can be generalized to QAM if the directional coupler splitter (DCS) before the balanced photodiode pair (BPD) in the Rx is replaced by a 90° hybrid. Delays applied in the pulse interleaver are multiples of $1/N\times$ the initial pulse repetition time and are exemplarilly shown as 0, $3 \times \tau_0$, $6 \times \tau_0$, and $9 \times \tau_0$, with τ_0 $1/12^{th}$ the pulse repetition time. The unmodulated reference pulse train is sent through the upper branch with a delay of $1 \times \tau_0$ that ensures that it is interleaved in between the modulated pulses before being combined with them, provided the effective number of comb lines exceeds $3 \times N$, i.e., 12 in this configuration, so that the pulses are sufficiently narrow not to overlap or so that they form orthogonal sinc-shaped pulse trains [50].

In the Rx, the combined signal is fed through an unbalanced interferometer that allows recovering every $1/N$ sample, i.e., that already implements deserialization in the optical domain. In the specific example shown in Fig. 3.16, this demodulator recovers the pulses that have been delayed by $3 \times \tau_0$. The other signal samples can be retrieved by splitting the light four-ways (N-ways) after in-coupling in the Rx PIC and sending it to 3 (resp. $N - 1$) more such demodulators, in which the number of delay loops in the top interferometer branch has been modified to 1, 1+3, and 1+6.

The functioning principle of the demodulator in Fig. 3.16b can be understood by considering the modulated pulse train to follow the upper branch of the interferometer and the reference pulse train to follow its low branch. The first delay τ_0 in the upper branch applies to the modulated pulses the same time delay already applied to the reference pulse in the transmitter. The following delays further shift the modulated pulses by multiples of $3 \times \tau_0$

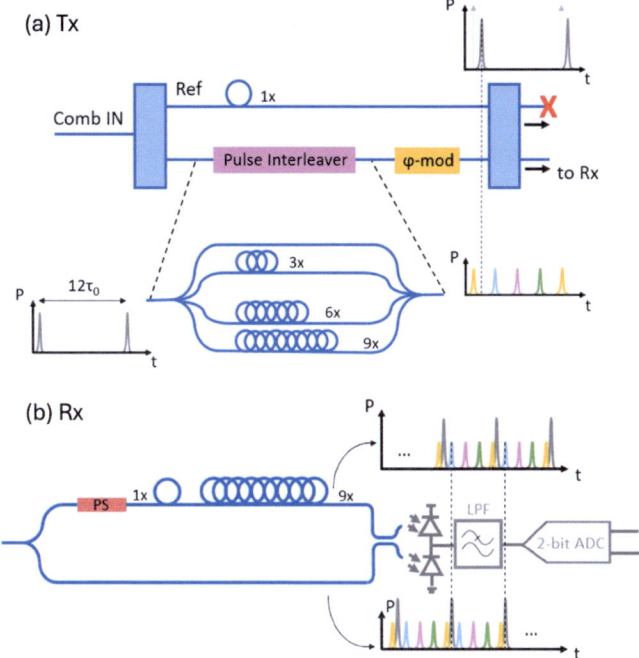

Fig. 3.16 Architecture of the optical link with built-in optical deserialization. **a** Tx and **b** Rx architecture. The time domain diagrams represent the initial pulse train (left diagram in (**a**)), as well as the pulse trains in the upper and lower branches of the Tx and Rx, respectively. Circles indicate optical delay loops in units of τ_0. (Reproduced under CC-BY-NC 4.0 from [7])

and select which of the N interleaved pulse trains are synchronized with the reference pulse and demodulated. The DCS at the end of the interferometer and the following BPD detect the phase difference between the modulated and reference pulse.

Of course, the modulated signal also follows the lower branch and the reference signal the upper branch. These paths do not, however, result in a differential signal at the output of the interferometer, as they result in timing in which the modulated and reference pulses continue to occupy different time slots. These do result, however, in a 3-dB optical power budget penalty, as the corresponding light is lost to signal demodulation. Other sources of power loss are the pulse interleaver in the Tx, as represented in Fig. 3.16a, but this is a simplified schematic used to facilitate the discussion. Since the pulse interleaver is not required to truly increase the FSR, as the interleaved pulses are not required to have constant phases, this operation can also be done in a lossless manner, for example with a lattice filter [7]. The combination of the modulated and reference pulse trains at the end of the Tx also results in 3-dB losses, but these are offset by the differential detection at the Rx and are the price to pay to transport both through the same fiber.

As a final processing step, the architecture relies on the ADC in the Tx to digitize the differential photocurrent after integrating it over one MLL pulse repetition time and could, for example, also have its reset function triggered by a properly delayed MLL pulse train, as descried in Sect. 3.2.1, to keep the system synchronous.

As for the scheme described in Sect. 3.3.2, this architecture does not actually require the pulses to be undispersed, and also works in the general case in which the phases between the comb lines are random but locked and slowly varying [6]. Another advantage of using the comb both as a carrier on which the signal is modulated and as a reference for demodulation, as opposed to using a single frequency carrier for the former, is that no DSP processing is required to compensate for these phases–this is also taken care of by the demodulator shown in Fig. 3.16b.

An extension of the receiver architecture consists in replacing the fixed delays in the upper branch, that are in excess of the first one, i.e., the 9 delay loops in Fig. 3.16b, by a programmable lattice filter [51] as shown in Fig. 3.17a. This device is configured as a programmable delay line (or programmable pulse shaper—PPS) that provides optical paths with delays 0, $3 \times \tau_0$, $6 \times \tau_0$ and $9 \times \tau_0$ with programmable phases and weights. This allows not only to select which of the interleaved pulse trains is demodulated, but also to detect a weighted superposition of the corresponding samples, which we have applied to equalization of inter-symbol interference (ISI) in a numerical model of the network.

Figure 3.17b and c show the result of 4-level pulse amplitude modulation (PAM4) with a high baud rate (200 GBd) and a bandwidth limited Tx (70 GHz driver and 80 GHz high-speed phase modulator, for example in reach of thin-film lithium niobate on insulator technology). The received signal samples are plotted as histograms. In Fig. 3.17b, the delays in the upper branches of the demodulators are assumed to be fixed as illustrated in Fig. 3.16 and set to detect a targeted sub-train of the interleaved pulses. It can be seen that the distributions in the histogram are heavily overlapping and the corresponding eye diagram thus closed. In Fig. 3.17c these delay lines are made programmable with a PPS implemented as a lattice filter, that is further trained to recover samples from a given sub-train with optimum equalization. In this case, four distinct distributions are again visible. This allows further reduction of the complexity of the following ADC, ideally to two bits ENOB, with some additional bits allocated to handling dynamic signal strength and offset. These simulations were performed in a comprehensive model comprising all relevant noise sources and assuming a comb power of 200 mW.

3.5 Conclusions

Photonically assisted analog-to-digital converters, that leverage the low jitter of best-of-class mode-locked lasers and break down broadband signals into multiple lower-speed tributaries, allow surpassing the performance limits of all-electronic data converters, increasing both

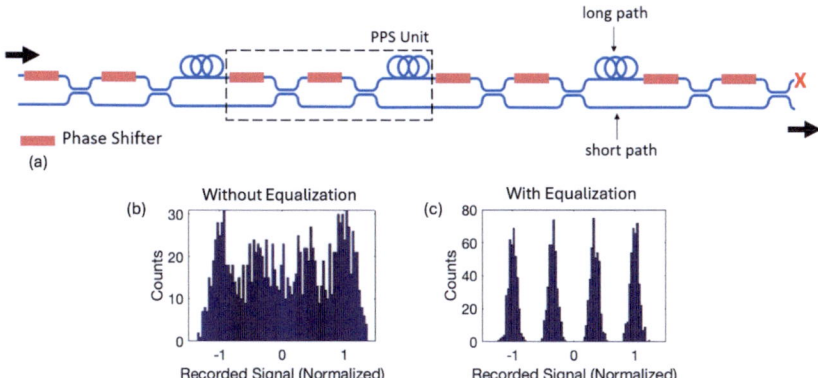

Fig. 3.17 a Architecture of a lattice filter used as a programmable delay line in the upper branch of the demodulator for dynamic channel reconfigurability and optical equalization. Histograms of the recovered signal samples with (**b**) fixed demodulator delays and (**c**) a trained lattice filter. (Reproduced under CC-BY-NC 4.0 from [7]

the ENOB by means of lower aperture jitter and increasing the optical signal bandwidth that is in reach of digitization.

We have reviewed our work on both time- and frequency-domain approaches and given an outlook on how these architectures can be further extended to include additional signal processing tasks such as equalization.

Optically triggered track-and-hold amplifiers with an equivalent jitter below 80 fs rms in a signal frequency range from 20 GHz to 70 GHz have been reported, which represent a significant advance over the current state-of-the-art.

Frequency-domain architectures implementing optical arbitrary waveform measurement up to signal bandwidths of 610 GHz have also been experimentally shown, with systems integrated at the chip-scale digitizing signals with bandwidths up to 160 GHz. The quality of recovered constellation diagrams has been benchmarked against that obtained from commercial broadband receivers and shown to surpass the latter. By combining this OAWM scheme with broadband electro-optic conversion by means of the thin-film lithium niobate modulator, a complete ADC with a record acquisition bandwidth of 320 GHz has been shown.

Extension of deserialization/digitization to further functionalities such as equalization can be done at little hardware overhead once optics have been introduced in the signal processing chain and open a path for further offloading of the electronic requirements. As baud rates continue to grow beyond 100 GBd in commercial systems, such signal processing schemes may increasingly come into focus to enable further scaling.

Acknowledgements The authors would like to acknowledge funding from the German Research Foundation (Deutsche Forschungsgemeinschaft—DFG) for project "Ultra-Wideband Photonically

Assisted Analog-to-Digital Converters" (403188360) in the framework for the Priority Programme 2111 "Electronic-Photonic Integrated Systems for Ultrafast Signal Processing" as well as from the German Federal Ministry for Research and Education for project "Scalable Photonic Neuromorphic Circuits", Grant 03ZU1106BA, in the framework of the Cluster4Future NeuroSys.

References

1. Kikuchi N, Hirai R, Fukui T (2015) Practical implementation of 100-Gbit/s/lambda optical short-reach transceiver with Nyquist PAM4 signaling using electroabsorptive modulated laser (EML). In: Proceedings of the optical fiber communication conference. Art. ID Th3A.2
2. Khilo A, Spector SJ, Grein ME, Nejadmalayeri AH, Holzwarth CW, Sander MY, Dahlem MS, Peng MY, Geis MW, DiLello NA, Yoon JU, Motamedi A, Orcutt JS, Wang JP, Sorace-Agaskar CM, Popović MA, Sun J, Zhou G-R, Byun H, Chen J, Hoyt JL, Smith HI, Ram RJ, Perrott M, Lyszczarz TM, Ippen EP, Kärtner FX (Feb2012) Photonic ADC: overcoming the bottleneck of electronic jitter. Opt Express 20:4454–4469
3. Benedick AJ, Fujimoto JG, Kärtner FX (2012) Optical flywheels with attosecond jitter. Nat Photonics 6:97–100
4. Zazzi A, Müller J, Gudyriev S, Marin-Palomo P, Fang D, Scheytt JC, Koos C, Witzens J (2020) Fundamental limitations of spectrally-sliced optically enabled data converters arising from MLL timing jitter. Opt Express 28(13):18790–18813
5. Haboucha A, Zhang W, Li T, Lours M, Luiten AN, Le Coq Y, Santarelli G (2011) Optical-fiber pulse rate multiplier for ultralow phase-noise signal generation. Opt Lett 36:3654–3656
6. Zazzi A, Das AD, Hüssen L, Negra R, Witzens J (2024) Scalable orthogonal delay-division multiplexed OEO artificial neural network trained for TI-ADC equalization. Photonics Res 12:85–105
7. Zazzi A, Witzens J (2024) Chip-to-chip orthogonal delay-division multiplexed network with optically enabled equalization. J Light Technol 42:7937–7953
8. Shinagawa M, Akazawa Y, Wakimoto T (1990) Jitter analysis of high-speed sampling systems. IEEE J Solid State Circuits 25(1):220–224
9. Kim J, Kärtner F (2010) Attosecond-precision ultrafast photonics. Laser & Photonics Rev 4(3):432–456
10. Bahmanian M, Scheytt JC (2021) A 2–20-GHz ultralow phase noise signal source using a microwave oscillator locked to a mode-locked laser. IEEE Trans Microw Theory Tech 69(3):1635–1645
11. Van Gasse K, Uvin S, Moskalenko V, Latkowski S, Roelkens G, Bente E, Kuyken B (2019) Recent advances in the photonic integration of mode-locked laser diodes. IEEE Photonics Technol Lett 31(23):1870–1873
12. Hauck J, Zazzi A, Merget F, Garreau A, Lelarge F, Moscoso-Mártir A, Witzens J (2019) Semiconductor laser mode locking stabilization with optical feedback from a silicon PIC. J Light Technol 37:3483–3494
13. Haus HA, Mecozzi A (1993) Noise of mode-locked lasers. J Quantum Electron 29:983–996
14. Kärtner FX, Singh N (2019) Integrated CMOS-compatible mode-locked lasers and their optoelectronic applications. In: 2019 IEEE BiCMOS and compound semiconductor integrated circuits and technology symposium (BCICTS)
15. Da Dalt N, Sheikholeslami A (2018) Understanding jitter and phase noise: a circuits and systems perspective. Cambridge University Press

16. Zazzi A, Müller J, Ghannam I, Battermann M, Vasudevan Rajeswari G, Weizel M, Scheytt JC, Jeremy W (2022) Wideband SiN pulse interleaver for optically-enabled analog-to-digital conversion: a device-to-system analysis with cyclic equalization. Opt Express 30:4444–4466
17. Weizel M, Kaertner FX, Witzens J, Scheytt JC (2020) Photonic analog-to-digital-converters–comparison of a MZM-sampler with an optoelectronic switched-emitter-follower sampler. In: 21st ITG-symposium on photonic networks, pp 119–124
18. Weizel M, Scheytt JC, Kärtner FX, Witzens J (2021) Optically clocked switched-emitter-follower THA in a photonic SiGe BiCMOS technology. Opt Express 29:16312–16322
19. Zazzi A, Müller J, Weizel M, Koch J, Fang D, Moscoso-Mártir A, Tabatabaei Mashayekh A, Das AD, Drayß D, Merget F, Kärtner FX, Pachnicke S, Koos C, Scheytt JC, Witzens J (2021) Optically enabled ADCs and application to optical communications. IEEE Open J Solid State Circuits Soc 1:209–221
20. Murmann B, ADC performance survey 1997–2024. [Online]. https://github.com/bmurmann/ADC-survey
21. Schwabe T, Kress C, Kruse S, Weizel M, Rhee H, Scheytt JC (2025) Forward-biased silicon phase shifter modeling for electronic-photonic co-simulation and validation in a 250 nm EPIC BiCMOS technology. J Light Technol 43(1):255–270
22. Bahmanian M, Fard S, Koppelmann B, Scheytt JC (2020) Wide-band frequency synthesizer with ultra-low phase noise using an optical clock source. In: 2020 IEEE/MTT-S international microwave symposium (IMS), pp 1283–1286
23. IEEE standard for terminology and test methods for analog-to-digital converters. IEEE Std 1241-2023 (Revision of IEEE Std 1241-2010) (2023)
24. Müller J, Zazzi A, Vasudevan Rajeswari G, Moscoso Mártir A, Tabatabaei Mashayekh A, Das AD, Merget F, Witzens J (2021) Optimized hourglass-shaped resonators for efficient thermal tuning of CROW filters with reduced crosstalk. In: Proceedings of the 17th IEEE international conference on group IV photonics (GFP)
25. Fontaine NK, Scott RP, Zhou L, Soares FM, Heritage JP, Yoo SJB (2010) Real-time full-field arbitrary optical waveform measurement. Nat Photonics 4(4):248–254
26. Solli DR, Ropers C, Koonath P, Jalali B (2007) Optical rogue waves. Nature 450(7172):1054–1057
27. Fontaine NK, Raybon G, Guan B, Adamiecki A, Winzer PJ, Ryf R, Konczykowska A, Jorge F, Dupuy J-Y, Buhl LL, Chandrashekhar S, Delbue R, Pupalaikis P, Sureka A (2012) 228-GHz coherent receiver using digital optical bandwidth interleaving and reception of 214-GBd (856-Gb/s) PDM-QPSK. In: European conference and exhibition on optical communication. Art. ID Th.3.A.1
28. Fang D, Zazzi A, Müller J, Drayss D, Füllner C, Marin-Palomo P, Mashayekh AT, Das AD, Weizel M, Gudyriev S, Freude W, Randel S, Scheytt JC, Witzens J, Koos C (2022) Optical arbitrary waveform measurement (OAWM) using silicon photonic slicing filters. J Light Technol 40(6):1705–1717
29. Drayss D, Fang D, Füllner C, Freude W, Randel S, Koos C (2024) Non-sliced optical arbitrary waveform measurement (OAWM) using a silicon photonic receiver chip. J Light Technol 42(14):4733–4750
30. Drayss D et al (2025a) Transmission of 300 GBd QAM signals over trans-oceanic distances using optical arbitrary waveform generation and measurement (OAWG/OAWM). J Light Techno 43(13) 6349–6360. https://doi.org/10.1109/JLT.2025.3579802
31. Drayss D, Fang D, Sherifaj A et al (2025b) Optical arbitrary waveform generation (OAWG) using actively phase-stabilized spectral stitching. Light Sci Appl 14:353. https://doi.org/10.1038/s41377-025-01937-4

32. Proietti R, Qin C, Guan B, Fontaine NK, Feng S, Castro A, Scott RP, Yoo SJB (2016) Elastic optical networking by dynamic optical arbitrary waveform generation and measurement. J Opt Commun Netw 8(7):A171–A179
33. Velasco L, Castro A, Asensio A, Ruiz M, Liu G, Qin C, Proietti R, Yoo SJB (2017) Meeting the requirements to deploy cloud RAN over optical networks. J Opt Commun Netw 9(3):B22–B32
34. Fang D, Drayss D, Peng H, Lihachev G, Füllner C, Kuzmin A, Marin-Palomo P, Kharel P, Wang RN, Riemensberger J, Zhang M, Witzens J, Scheytt JC, Freude W, Randel S, Kippenberg TJ, Koos C (2025) 320 GHz photonic-electronic analogue-to-digital converter (ADC) exploiting Kerr soliton microcombs. Light Sci & Appl 14. Art. ID 241
35. Drayss D, Fang D, Füllner C, Lihachev G, Henauer T, Chen Y, Peng H, Marin-Palomo P, Zwick T, Freude W, Kippenberg TJ, Randel S, Koos C (2023) Non-sliced optical arbitrary waveform measurement (OAWM) using soliton microcombs. Optica 10(7):888–896
36. Chen X, Xie X, Kim I, Li G, Zhang H, Zhou B (2009) Coherent detection using optical time-domain sampling. IEEE Photonics Technol Lett 21(5):286–288
37. Fischer JK, Ludwig R, Molle L, Schmidt-Langhorst C, Leonhardt CC, Matiss A, Schubert C (2011) High-speed digital coherent receiver based on parallel optical sampling. J Light Technol 29(4):378–385
38. Drayss D, Fang D, Füllner C, Likhachev G, Henauer T, Chen Y, Peng H, Marin-Palomo P, Zwick T, Freude W, Kippenberg TJ, Randel S, Koos C (2022) Slice-less optical arbitrary waveform measurement (OAWM) in a bandwidth of more than 600 GHz. In: Optical fiber communications conference. Art. ID M2I.1
39. Drayss D, Fang D, Füllner C, Kuzmin A, Freude W, Randel S, Koos C (2022) Slice-less optical arbitrary waveform measurement (OAWM) on a silicon photonic chip. In: European conference on optical communication. Art. ID We4E.6
40. Nguyen RL, Mellati A, Fernandez A, Iyer A, Fan A, Reyes B, Abidin C, Nani C, Albano D, Ahmad F, Solis F, Minoia G, Hatcher G, Bachu M, Garampazzi M, Hassanpourghadi M, Fan N, Prabha P, Fan S, Ho S, Dusatko T, Wu T, Elsharkasy W, Sun Z, Jantzi S, Tse L (2024) 18.4 A 200GS/s 8b 20fJ/c-s receiver with >60GHz AFE bandwidth for 800Gb/s optical coherent communications in 5nm FinFET. In: IEEE international solid-state circuits conference, pp 344–346
41. Infiniium UXR-series oscilloscopes data sheets. [Online]. https://www.keysight.com/us/en/assets/7018-06242/data-sheets/5992-3132.pdf
42. Deakin C, Liu Z (2022) Frequency interleaving dual comb photonic ADC with 7 bits ENOB up to 40 GHz. In: Conference on lasers and electro-optics (CLEO). Art. ID STh5M.1
43. Valley GC (2007) Photonic analog-to-digital converters. Opt Express 15(5):1955–1982
44. Herr T, Brasch V, Jost JD, Wang CY, Kondratiev NM, Gorodetsky ML, Kippenberg TJ (2014) Temporal solitons in optical microresonators. Nat Photonics 8(2):145–152
45. Kippenberg TJ, Gaeta AL, Lipson M, Gorodetsky ML (2018) Dissipative kerr solitons in optical microresonators. Science 361(6402). Art. ID eaan8083
46. Shen B, Chang L, Liu J, Wang H, Yang Q-F, Xiang C, Wang RN, He J, Liu T, Xie W, Guo J, Kinghorn D, Wu L, Ji Q-X, Kippenberg TJ, Vahala K, Bowers JE (2020) Integrated turnkey soliton microcombs. Nature 582(7812):365–369
47. Mercante AJ, Shi S, Yao P, Xie L, Weikle RM, Prather DW (2018) Thin film lithium niobate electro-optic modulator with terahertz operating bandwidth. Opt Express 26(11):14810–14816
48. Zhang Y, Shao L, Yang J, Chen Z, Zhang K, Shum K-M, Zhu D, Chan CH, Lončar M, Wang C (2022) Systematic investigation of millimeter-wave optic modulation performance in thin-film lithium niobate. Photonics Res 10(10):2380–2387
49. Liu J, Lucas E, Raja AS, He J, Riemensberger J, Wang RN, Karpov M, Guo H, Bouchand R, Kippenberg TJ (2020) Photonic microwave generation in the X- and K-band using integrated soliton microcombs. Nat Photonics 14(8):486–491

50. Hosni MI, Meier J, Mandalawi Y, Singh K, Mandal P, Elghandur AH, Schneider T (2023) Orthogonal sampling-based broad-band signal generation with low-bandwidth electronics. IEEE Open J Commun Soc 4:2930–2938
51. Bohn M, Rosenkranz W, Krummrich PM (2004) Adaptive distortion compensation with integrated optical finite impulse response filters in high bitrate optical communication systems. J Sel Top Quantum Electron 10:273–280

Open Access This chapter is licensed under the terms of the Creative Commons Attribution 4.0 International License (http://creativecommons.org/licenses/by/4.0/), which permits use, sharing, adaptation, distribution and reproduction in any medium or format, as long as you give appropriate credit to the original author(s) and the source, provide a link to the Creative Commons license and indicate if changes were made.

The images or other third party material in this chapter are included in the chapter's Creative Commons license, unless indicated otherwise in a credit line to the material. If material is not included in the chapter's Creative Commons license and your intended use is not permitted by statutory regulation or exceeds the permitted use, you will need to obtain permission directly from the copyright holder.

Precise Optical Nyquist Pulse Synthesizer Digital-to-Analog Converter

J. Christoph Scheytt, Tobias Schwabe, Karanveer Singh, Christian Kress and Thomas Schneider

Abstract

Optically assisted digital-to-analog converters (DACs) using Nyquist pulse sequences (NPSs) are presented and investigated. Therefore, NPSs are mathematically described and analyzed. Based on this, the operating principle of a precise optical Nyquist pulse synthesizer digital-to-analog converter (PONyDAC) is described. Possible architectures of PONyDAC are derived and compared in terms of performance and practicability. Moreover, the limits of PONyDAC systems and their superiority over classical electronic DACs are discussed. Furthermore, discrete building-block based implementations and monolithic implementations in electronic-photonic integrated circuits (EPICs) are presented. To enable a practicable monolithic integration, a shrinkage of the Mach-Zehnder modulators (MZMs) has been performed by applying forward-biased phase shifters (FB-PSs). These FB-PSs are analyzed and modeled to allow the precise and reliable design of PONyDAC systems with multiple MZMs. Finally, data conversion and data transmission experiments are carried out to demonstrate the systems functionality, quantify its performance, and prove their superiority over purely electronic DACs.

Acronyms

AWG	Arbitrary Waveform Generator
BiCMOS	Bipolar and Complementary Metal-Oxide-Semiconductor Transistor
BW	Bandwidth
CMOS	Complementary Metal-Oxide-Semiconductor Transistor
CW	Continuous-Wave
DAC	Digital-to-Analog Converter
DT-PS	Depletion-Type Phase Shifter
EF	Emitter Follower
ENOB	Effective Number of Bits
ET-PS	Enhancement-Type Phase Shifter
ER	Extinction Ratio
EPIC	Electronic-Photonic Integrated Circuit
FB-PS	Forward-Biased Phase Shifter
FCPD	Free Carrier Plasma Dispersion Effect
$LiNbO_3$	Lithium Niobate
MZM	Mach-Zehnder Modulator
NPS	Nyquist Pulse Sequence
NRZ	Non-Return-to-Zero
OFC	Optical Frequency Comb
OSA	Optical Spectrum Analyzer
PAM	Pulse-Amplitude-Modulation
PONYDAC	Precise Optical Nyquist Pulse Synthesizer Digital-to-Analog Converter
PRBS	Pseudorandom Binary Sequence
PS	Phase Shifter
RIN	Relative Intensity Noise
SI	Silicon
SINAD	Signal-to-Noise-and-Distortion Ratio
SCR	Space-Charge-Region
SD	Segment Driver
SNR	Signal-to-Noise Ratio
TL	Transmission Line
TWE	Traveling-Wave Electrode

4.1 Introduction

DACs are important components bridging between the digital and analog domains, enabling the conversion of discrete digital signals into continuous analog signals. They are essential in a variety of systems, such as audio and video processing, digital control circuits, measurement and communication systems. As the demands on modern electronic systems increase in all these application fields, so does the need for ever more powerful and precise converters. DACs are often designed at the limit of the respective applied technology, but higher precisions are still desired and required. To meet these requirements, photonically assisted DACs, such as the PONyDAC, can be used, as they enable system bandwidths (BWs) that are multiple times higher than the BW of the applied electrical and elctro-optical components. Using optical NPSs, multiple digital input signals can be modulated onto an optical carrier in different channels and combined afterwards, resulting into very high BW and precision.

In the following, the PONyDAC system is presented and analyzed. First, the fundamental operational principle of the PONyDAC is explained and an overview on possible topologies is given. The advantages and disadvantages of possible architectures are discussed and evaluated. Afterwards, EPIC implementations of the PONyDAC channels are presented and different applicable modulator types are compared. Experimental results are presented, which proof the overall system functionality and applicability. Moreover, theoretical works, based on those results, are presented, which proof, that the PONyDAC system is able to overcome limits of today's state-of-the-art electrical DACs. Finally, a conclusion is given, which summarizes all investigations and results.

4.2 Topologies of Electro-Optical DACs

The sampling theorem states that a BW-limited signal can be expressed as a sum of time-shifted sinc-pulses, each of which weighted with the values of the sampling points [1]. Due to the orthogonality between the sinc-pulses, any subsequent time-shifted pulse must coincide with the zero crossings of all previous pulses. This observation is the starting point of the PONyDAC proposed in this chapter. Since sinc pulses, whose frequency spectrum is a rectangular function (see Fig. 4.1b), are unlimited in the time domain, they are just a mathematical construct and not realizable in practice.

Therefore, PONyDACs use NPSs, which are sinc-pulse sequences instead of sinc-pulses (see Fig. 4.1c). However, it can be shown mathematically that NPSs, whose frequency spectrum is a comb with N lines (see Fig. 4.1d), can be described as an infinite sum of ideal sinc-pulses, with each pulse having a particular time shift [2]:

$$s_{NPS}(t) = \frac{\sin(\pi N \Delta f t)}{N \cdot \sin(\pi \Delta f t)} = \sum_{n=-\infty}^{\infty} (-1)^{(N-1)\cdot n} \cdot \mathrm{sinc}\left(N \cdot \Delta f \cdot \left(t - \frac{n}{\Delta f}\right)\right) \quad (4.1)$$

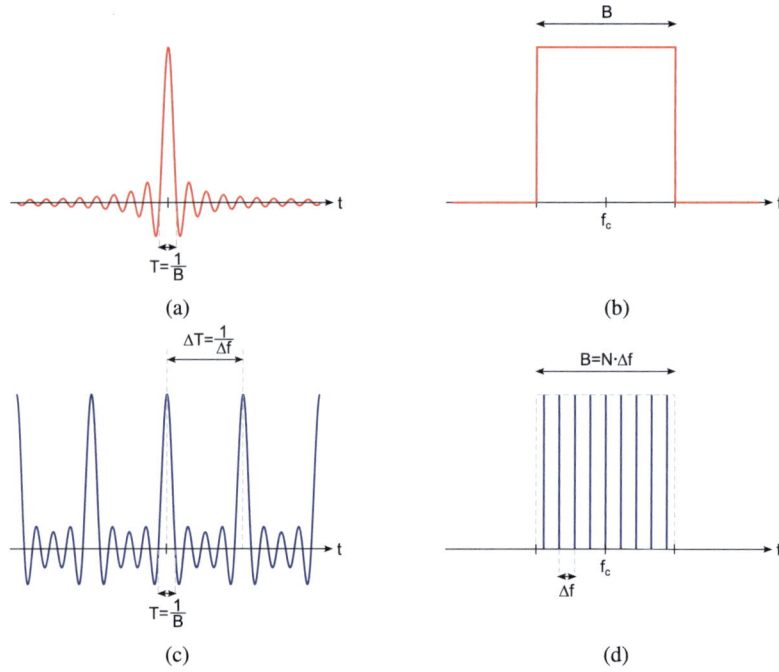

Fig. 4.1 **a** Nyquist pulse in time domain **b** Nyquist pulse frequency spectrum **c** NPS in time domain ($N = 9$) **d** NPS frequency spectrum ($N = 9$)

$$S_{NPS}(f) = \mathcal{FT}\left(\frac{\sin(\pi N \Delta f t)}{N \cdot \sin(\pi \Delta f t)}\right) = \sum_{n=-\frac{N-1}{2}}^{\frac{N-1}{2}} \delta(f - n \cdot \Delta f). \quad (4.2)$$

Thus, the superposition of NPSs is also a superposition of time-shifted sinc pulses. Therefore, orthogonal NPSs can also describe BW-limited signals without any errors.

As presented in Fig. 4.2, the PONyDAC consists of three parts. In a first step, NPSs are generated in multiple channels. To ensure orthogonality and fault-free superposition in the last building block, these NPSs are delayed against each other. Afterwards these pulse sequences are weighted according to electronic digital input signals within multiple

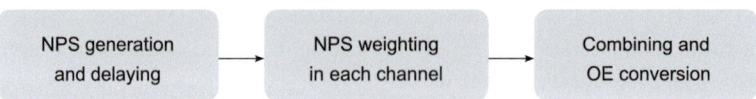

Fig. 4.2 PONyDAC building blocks

channels. At last, all channels' signals are combined and the optical signal is converted to the electronic domain.

Within the different PONyDAC topologies, presented in the following Subsections, only the first building block differs. In both presented topologies, a rectangular optical frequency comb (OFC) is generated in the first building block. This OFC corresponds to the sinc-shaped NPS in the time domain, as illustrated in Fig. 4.1d for an NPS with nine frequency lines ($N = 9$). A significant advantage of sinc-pulse sequences is that they can be generated by one or more cascaded modulators, offering control over BW and repetition rates while supporting integration within a photonic platform. This technique yields NPSs of superior quality, as the resulting OFC can be ideally flat and rectangular. Moreover, the generated pulse BW can reach three to four times the electro-optical BW of the modulators used [2].

The generation of sinc-pulse sequences with modulators provides a crucial advantage, as this pulse synthesizer modulator can be integrated into a Silicon (Si) photonic platform, making the complete Photonic DAC concept complementary metal-oxide-semiconductor transistor (CMOS) compatible. The first experimental demonstration of sinc-pulse sequence generation was presented in [2], with bench-top Lithium Niobate ($LiNbO_3$) modulators. In [3], an integrated MZM generated sinc-pulse sequences for optical sampling applications. In a follow-up study, the BW bottleneck of the modulators was overcome by driving a 19 GHz 3 dB-BW integrated modulator with a 30 GHz signal to generate high-quality sinc-pulse sequences of 90 GHz BW [4].

Once the frequency synthesis has been established, the PONyDAC system can be implemented in two different ways, for example in an available Si photonics platform. The first option is to apply a single pulse generator and use optical delay lines for pulse alignment, named Photonic DAC with Single Pulse Synthesizer, and the second option applies multiple NPS generators aligned with electrical tuning and termed as Photonic DAC with Multiple Pulse Synthesizers. In the following Section, the complete working principles of the two implementations and their advantages are discussed.

4.2.1 Photonic DAC with Single Pulse Synthesizer

The photonic DAC concept with a single pulse source is shown in Fig. 4.3 for three channels. A continuous-wave (CW) laser output is modulated with a single-tone sinusoidal signal of frequency f_m by a single MZM, and the operating bias point is selected such that the output optical spectrum with central frequency f_c and sidebands at $f_c - f_m$ and $f_c + f_m$ have equal power levels. This rectangular three-line comb corresponds to an NPS with two zero crossings. Thus, the modulation frequency f_m defines the repetition rate of the peaks inside the frequency comb and the corresponding repetition period $T = 1/f_m$ between the sinc pulses. For a larger number of channels, and therefore also higher bandwidths, additional MZMs can be added behind the first MZM. This allows an increase of lines in the OFC and further channels can be added to increase the system's overall output sampling rate T_S.

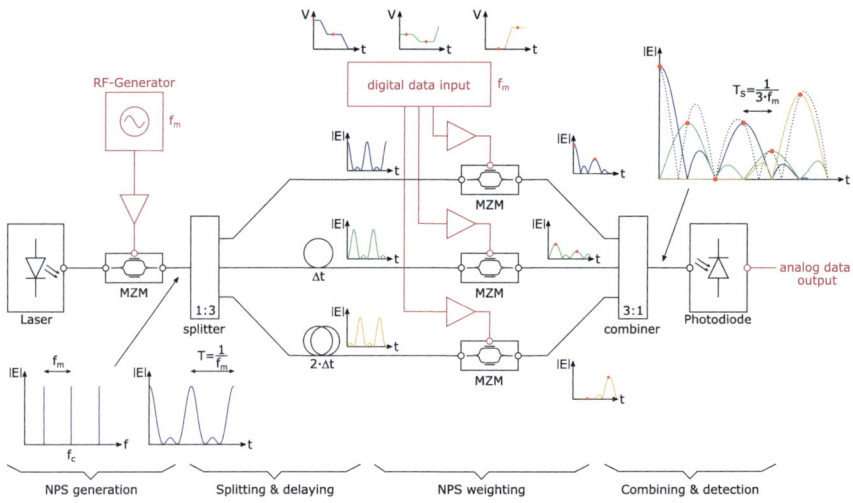

Fig. 4.3 Three channel photonic DAC with Single Pulse Synthesizer

Afterwards, the output of the first modulator is split into multiple channels. Since the proposed concept relies on the orthogonality of the generated NPSs in each channel, the pulses need to be aligned orthogonal to each other. For the pulse alignment, optical delay lines with specific individual delays for each channel need to be incorporated. The input of each channel has to be delayed by $360°/N$ to each other, which corresponds to a time delay of $\Delta t = \frac{1}{N \cdot f_m}$ with N being the amount of channels and f_m the modulation frequency applied to the input MZMs in the NPSs synthesis. Thus, in the three-channel ($N = 3$) example shown in Fig. 4.3, the first channel with no optical delay ($\Delta t_1 = 0$) acts as a reference point for the other channels, which require time delays of $\Delta t_2 = \frac{1}{3} \cdot \frac{1}{f_m}$ and $\Delta t_3 = \frac{2}{3} \cdot \frac{1}{f_m}$, respectively. Depending on the comb spacing, as defined by the NPSs synthesis frequency f_m, these delay lines with specific delays and thermal tuning need to be designed. This is one of the major limitations on the acceptable range of the frequency f_m, which directly limits the overall sampling rate of the Photonic DAC.

In all channels, the delayed pulses align orthogonal to each other, which means that the peak of the NPSs in one channel aligns with the zero crossing of the pulses in the other channels. This orthogonality plays an important role in preserving each data point of the resulting signal generated with the Photonic DAC.

In a next step, the NPSs are weighted according to electronic digital input data. Therefore, low-BW electronic DACs generate data signals, which are fed to the MZMs in each channel. Each of these signals carries $1/N$ of the information of the targeted analog output waveform. By means of additional MZMs, the data points are transferred into the optical domain, defining the peak of the NPS in each channel. Afterwards, the sampled NPSs are combined and converted into electronic domain using a photodiode. Thereby, a high BW signal with a sampling rate of N times the electronic BW of the MZMs in each channel is generated. For example, in a three-channel ($N = 3$) system the pulses with a repetition rate f_m are weighted by data signals with a sampling rate of f_m, resulting in an overall signal of sampling rate $f_S = 3 \cdot f_m$.

The described prototype employs CMOS-compatible components that can be easily integrated into any Si photonic platform. The tuning range and achievable BW of the system are defined by the generated pulses and the optical delay lines. At higher BWs, it is quite challenging to have precisely aligned optical delays with on-chip delay lines, and they are more sensitive to thermal crosstalk. However, this type of implementation has the advantage that the number of modulators required only increases with $N + \log_3(N)$ when the number of channels N is increased. In the next Section, the concept is upgraded with electronically tunable delays, which provide more flexibility and tunability for the system.

4.2.2 Photonic DAC with Multiple Pulse Synthesizers

A PONyDAC system with three channels and multiple pulse synthesis is shown in Fig. 4.4. The overall design consists of $N \cdot (1 + \log_3(N))$ MZMs, leading to two more MZMs in a three-channel system than in the previous design. However, no optical delay lines are required and the optical alignment of the pulses is carried out with electrical phase shifters (PSs), allowing flexible tuning of phase shifts and therefore also flexible modulation frequencies f_m. The laser output is split into multiple channels, each of them consisting of MZMs for NPS generation and optical alignment and an MZM to weight the pulses. The pulse alignment is done via phase shifts between the electrical input signals of the NPS generation MZMs. Therefore, the required electrical PSs can easily be integrated in EPICs in a tunable

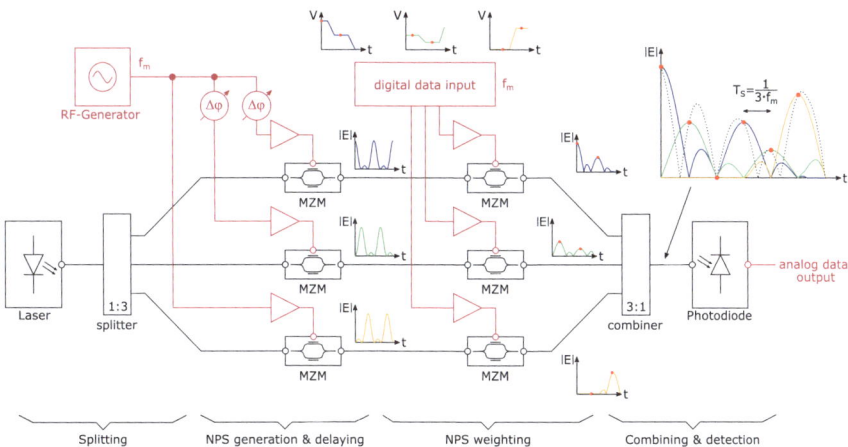

Fig. 4.4 Three channel photonic DAC with Multiple Pulse Synthesizers

and area-efficient manner. This provides a great advantage in terms of range and tunability of the output frequency of the generated signals.

4.3 EPIC Implementations in Silicon Technology

Monolithic implementations enable more compact solutions with less electrical parasitics, compared to hybrid implementations. Furthermore, they allow cost-efficient mass-production with lower tolerances and high reliability. With regards to PONyDAC systems, compactness and reduced tolerances are important, since the system strongly relies on precise electrical and optical delays and contains many electronic-photonic interfaces (see Sect. 4.2), which are difficult to handle in hybrid solutions. In addition to different architectures (described in Sect. 4.2), PONyDAC systems can also differ in terms of applied modulator types, which strongly effects the modulator driver structure and the overall system's size. Both investigated MZM types, which are described in the following Subsections, include a splitter, thermal PSs for MZM biasing, a combiner and free carrier plasma dispersion effect (FCPD) PSs, as depicted in Fig. 4.5a.

According to the FCPD, the amount of free charge carriers inside the Si effects the silicon's refractive index and absorption coefficient [5]:

$$\Delta n = -\frac{e^2 \lambda^2}{8\pi^2 c^2 \epsilon_0 n} \left(\frac{\Delta N_e}{m^*_{ce}} + \frac{\Delta N_h}{m^*_{ch}} \right) \tag{4.3}$$

$$\Delta \alpha = -\frac{e^3 \lambda^2}{4\pi^2 c^3 \epsilon_0 n} \left(\frac{\Delta N_e}{m^{*2}_{ce} \mu_e} + \frac{\Delta N_h}{m^{*2}_{ch} \mu_h} \right) \tag{4.4}$$

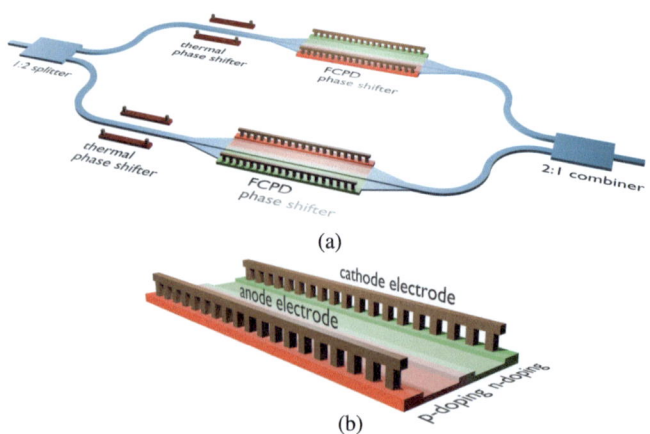

Fig. 4.5 **a** MZM and **b** FCPD Si PS geometry

with refractive index n, absorption coefficient α, elementary charge e, optical wavelength λ, speed of light in vacuum c, vacuum permittivity ϵ_0, mobile charge carrier concentrations ΔN_e and ΔN_h, effective masses m_{ce}^* and m_{ch}^*, and mobilities μ_e and μ_h of electrons and holes, respectively.

With P-N-dopings, a diode structure is built in a Si waveguide to form the FCPD based PS, as illustrated in Fig. 4.5b. Assuming an optical input signal

$$E_{in}(t,z) = E_0 \cdot e^{j\omega t} \cdot e^{-j\frac{2\pi}{\lambda_0} n_{eff} z} \cdot e^{-\alpha z} \tag{4.5}$$

entering the PS, an effective refractive index change $\Delta n_{eff}(v_{PS})$ and an attenuation coefficient change $\Delta \alpha_{eff}(v_{PS})$ in the waveguide, caused by an applied PS voltage v_{PS}, lead to an output signal [6]

$$E_{out}(t,l,v_{PS}) = E_0 \cdot e^{j\omega t} \cdot e^{-j\phi_0} \cdot e^{-l\alpha_0} \cdot e^{-j\Delta\phi(v_{PS})} \cdot e^{-l\Delta\alpha(v_{PS})} \tag{4.6}$$

with the static and the dynamic, electrically controlled phase shifts

$$\phi_0 = \frac{2\pi}{\lambda_0} \cdot n_{eff,0} \cdot l \quad and \quad \Delta\phi(v_{PS}) = \frac{2\pi}{\lambda_0} \cdot \Delta n_{eff}(v_{PS}) \cdot l \tag{4.7}$$

and with the time t, PS length l, input field amplitude $E_0 = E_{in}(t=0, z=0)$, angular frequency ω, vacuum wavelength λ_0, effective refractive index $n_{eff,0}$, and attenuation coefficient α_0 at $v_{PS} = 0$, respectively.

Therefrom, under the assumption of ideal optical splitters, lossless thermal PSs adding a phase shift of ϕ_{th}, ideal combiners and negligibly small attenuation coefficient change $\Delta\alpha_{eff}(v_{PS})$, the overall MZM transfer function follows as:

$$A_{MZM} = \frac{P_{out}}{P_{in}} \approx e^{-2l\alpha_0} \cdot \cos^2(\frac{1}{2} \cdot \Delta\phi_{diff} + \frac{1}{2} \cdot \phi_{th}) \tag{4.8}$$

with the differential phase shift $\Delta\phi_{diff} = \Delta\phi_{PS,1} - \Delta\phi_{PS,2}$.

4.3.1 EPIC Implementations with Depletion-Type Modulators

Depletion-type phase shifters (DT-PSs) are widely used in modern Si photonic electro-optical transmitters [7]. In DT-PS, the pn-junction is operated in reverse biased condition ($v_{PS} < 0$) and a space-charge-region (SCR) with voltage dependent width $w_{SCR}(v_{PS})$ is built (see Fig. 4.6a). Inside the SCR, the Si is depleted, leading to reduced free charge carriers interacting with the electrical field of the optical signal. Thus, according to Eqs. 4.3 and 4.7, the phase shift $\Delta\phi(n_{eff})$ is electrically controlled by the applied PS voltage v_{PS}.

Assuming a negligibly small reverse leakage current, the DT-PS can be modeled electrically as an RC series network, as depicted in Fig. 4.6b [8]. The resistance R_S is given by the contact resistance and the ohmic resistance of the semiconductor material, while the junction capacitance c_J follows from the SCR which acts as a depletion layer between charges in the P and the N doped regions. It is given as [9]:

$$c_j(v_J) = \frac{C_{J0}}{(1 - \frac{v_J}{\phi_J})^M} \tag{4.9}$$

with $C_{J0} = c_j(0)$, ϕ_J and M being the zero-voltage junction capacitance, the built-in junction potential and the ideality factor, respectively.
As the junction capacitor voltage v_J is causing the refractive index change, the intrinsic electrical 3 dB-PS-BW follows as:

$$f_{-3dB,DT-PS,el} = \frac{1}{2\pi \cdot R_S \cdot c_j}. \tag{4.10}$$

Compared to other FCPD PS, DT-PS have a comparably high BW, but low efficiency. In consequence, a long PS length, which is usually in the mm-range is required to achieve sufficient phase shift swings. Therefore, the electrical wavelength $\lambda_{el} = c_0/f_{el}$ exceeds the DT-PS length for frequencies above 300 MHz and transmission lines (TLs) between the PSs and the driver circuitry are required to allow proper signal distribution. Moreover, velocity matching between the electrical signal in the TL and the optical signal in the PS is required to avoid increased microwave losses and reduced BW. The overall phase transfer function is given as [10]:

$$H(f) = e^{\frac{\alpha L}{2}} \sqrt{\frac{\sinh^2(\frac{\alpha L}{2}) + \sinh^2(\frac{\Psi L}{2})}{(\frac{\alpha L}{2})^2 + (\frac{\Psi L}{2})^2}} \tag{4.11}$$

with $\alpha(f)$ describing the microwave loss, L being the PS length, $\Psi(f) = 2\pi \frac{f}{c_0}(n_{\text{eff}}^e - n_{\text{eff}}^o)$, n_{eff}^e being the electrical and n_{eff}^o the optical effective refractive group index.

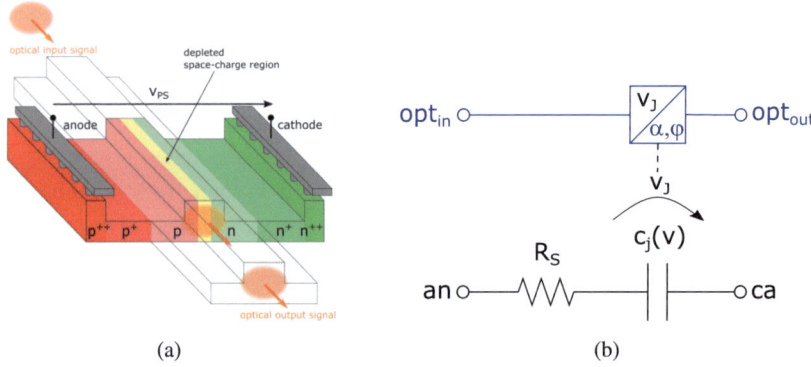

Fig. 4.6 DT-PS **a** illustration with SCR **b** small-signal equivalent circuit

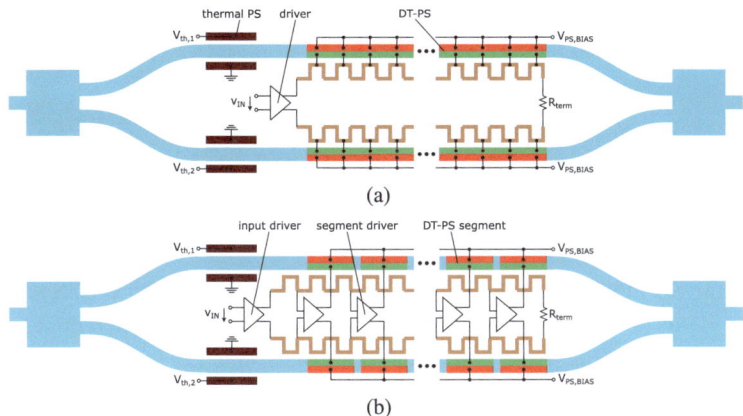

Fig. 4.7 Depletion-type MZM driving schemes **a** directly driven TWE **b** SD

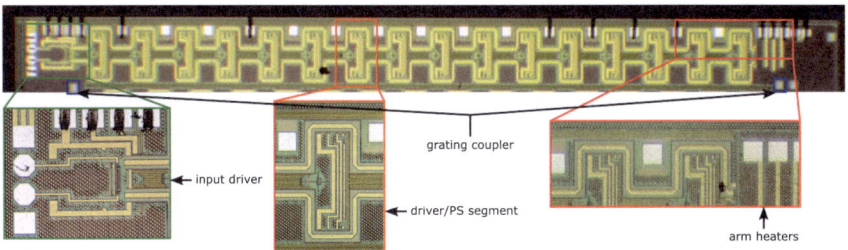

Fig. 4.8 Chip picture of the fabricated EPIC MZM [8]

Velocity matching can be achieved by careful TL design with n_{eff}^e adaption, which is, due to technology restrictions with given metal stacks, only possible to a limited extent, or by meandering the TL, which is effectively decreasing the electrical velocity.

Two different driving schemes can be applied to connect the PS driver, the TL and the DT-PS. In the directly driven traveling-wave electrode (TWE) scheme, which is shown in Fig. 4.7a, the TL is directly connected to the DT-PS, which causes significant drawbacks, because the TL is directly loaded by the DT-PS capacitance. Therefrom, the characteristic impedance $Z_0 = \sqrt{L_0/(C_0 + C_L)}$ and the electrical propagation delay $\tau_e = \sqrt{L_0 \cdot (C_0 + C_L)}$ are strongly effected [11]. Since the capacitive load is nonlinear (see Eq. 4.9) and changes during operation, permanent conjugate load matching between the driver and the TL is not possible, leading to reduced efficiency. Additionally, the large load capacitance causes microwave losses and reduces the BW [8].

Another approach, which is presented in Fig. 4.7b, is the segmented driver scheme. In the segmented driver scheme, the DT-PS is split into short elements, each individually driven by a segment driver (SD), which is connected to the TL. Thus, the TL load capacitance is

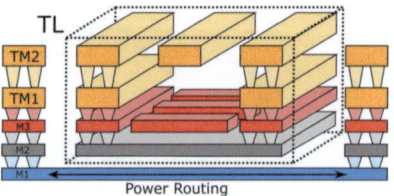

Fig. 4.9 Cross section of the slow-wave TL [8]

reduced to the SD input impedance. Because of these advantages, the DT-PS based MZMs in the PONyDAC demonstrators have been implemented with segmented drivers [8, 12].

Figure 4.8 shows a chip photo of a designed and measured DT-PS based MZM [8]. The entire chip contains two of the presented MZMs, representing a channel of the PONyDAC system shown in Fig. 4.4 and is implemented in a 250 nm bipolar and complementary metal-oxide-semiconductor transistor (BiCMOS) technology [13]. It is a further development of a previously published MZM [14], with improved TL design, which is illustrated in Fig. 4.9. As illustrated, the signal line on the highest top metal (TM in Fig. 4.9) is shielded by a four metal layer stack (M2 to TM2 in Fig. 4.9) ground line, allowing excellent shielding from three sides. Additional floating metal structures (on level M3 in Fig. 4.9) between the signal line and the ground line lead to reduced microwave speed and thereby to better velocity matching between the optical and the electrical signal at reduced TL length and thereby reduced microwave loss. Moreover, the TL is designed to $Z_T = 40\,\Omega$ characteristic impedance, and therefore it is terminated by a 40 Ω resistor, as well. Decreasing the characteristic impedance Z_T leads to increased signal linewidth, reduced inductance and thereby to increased BW, but it also increases the TL size and thereby the capacitance. Additionally, it increases the power dissipation in the input driver. Considering all these effects and limited chip area for the TL, 40 Ω was found as a good trade-off [8].

To match the TL's characteristic impedance, a differential cascode amplifier with $R_C = 40\,\Omega$ collector resistance is used in the input driver (shown in Fig. 4.10a). The cascode topology offers high voltage gain and reduces the Miller effect, leading to increased BW [15]. In addition, emitter degeneration and capacitive peaking are applied by means of the resistor R_E and the capacitor C_E in Fig. 4.10a, to enhance the BW further. The cascode amplifier follows an emitter follower (EF) input stage with 50 Ω input resistors to match high-frequency measurement equipment standard output resistance. The EF provides high input resistance allowing good matching by the 50 Ω resistors, high BW and low output resistance to drive the cascode amplifier [16]. All stages are biased by current mirror current sources offering precise and adjustable currents at high output resistance.

In same manner, the SD is designed. As illustrated in Fig. 4.10b, an EF input stage with low output resistance and high input resistance to reduce loading of the TL, drives a capacitively peaked cascode amplifier. In addition, capacitors C_{comp} are inserted to compensate the input referred capacitance of the SD. Moreover, EF output stages are added to increase the driving capability of the SD and reduce the BW reduction due to the PS capacitance.

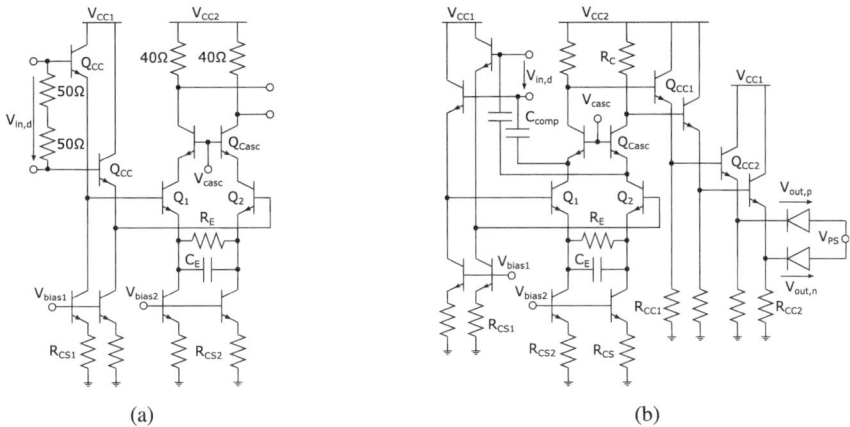

Fig. 4.10 Circuits of the implemented DT-PS based electro-optical transmitter [8] **a** input driver **b** SD

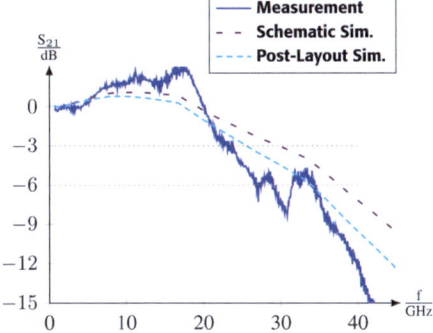

Fig. 4.11 Measured and simulated S21 electro-optical response of EPIC MZM with external 70 GHz photodiode [8]

The electro-optical transmitter (MZM and driver) is operating in the optical C-band ($\lambda_{opt} = 1550$ nm) and exhibits an electro-optical 3 dB-BW of $f_{-3dB} = 24$ GHz, a 6 dB-BW of $f_{-6dB} = 34$ GHz, a 10 dB extinction ratio (ER) at 30 Gb/s, and an input-referred π-phase shift voltage of $V_\pi = 420$ mV at a power consumption of 1.7 W and a footprint of 5.06 mm^2 (Fig. 4.11).

It allows data transmission with data rates up to 60 Gb/s in non-return-to-zero (NRZ) scheme and up to 80 Gb/s in 4-level pulse-amplitude-modulation (PAM) scheme, as shown in Fig. 4.12.

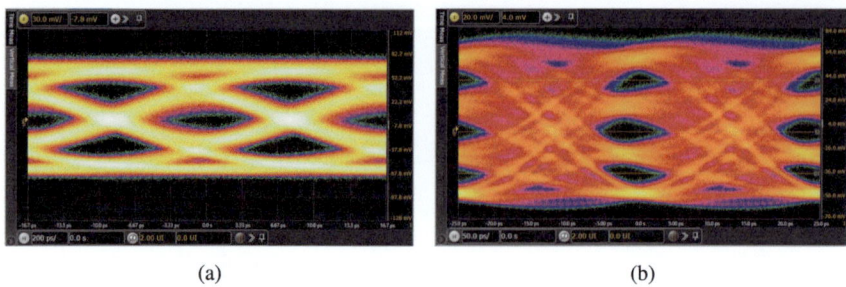

Fig. 4.12 Measured eye diagrams in data transmission experiment [8] **a** 60 Gb/s NRZ **b** 80 Gb/s PAM-4

4.3.2 EPIC Implementations with Enhancement-Type Modulators

In contrast to the DT-PS described in the previous Subsection, enhancement-type phase shifters (ET-PSs) are less frequently used due to their significantly lower BW, when using no additional equalizing circuitry. Nevertheless, ET-PSs exhibit a very large modulation efficiency and and equalizer circuits can be applied to increase the electro-optical BW [17, 18]. Therefore, area-efficient high-speed MZMs can be realized with short ET-PS length. In ET-PSs, the pn-junction is biased in forward-direction, leading to a different equivalent circuit, when exceeding the built-in potential $v_D > \Phi_J$, as depicted in Fig. 4.13.

The PS pn-junction is modeled by the ideal diode D, which is following the Shockley equation. As no SCR is built in forward-bias operation, no junction capacitance is present, but a diffusion capacitance c_d is given, which is caused by excess charge carriers before recombination [9]. The inductor L_S is modeling the contact inductance, while R_S is modeling the ohmic behavior of the semiconductor and the contact metallization. Additionally, the resistance and capacitance of the anode and cathode contacts are taken into account by the resistors and capacitors R_{an}, R_{ca}, C_{an}, and C_{ca}, respectively. Besides the electrical

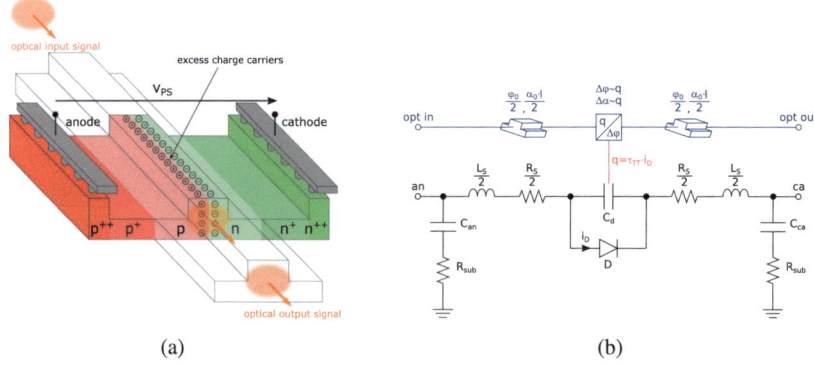

Fig. 4.13 ET-PS **a** illustration with SCR **b** equivalent circuit [6]

4 Precise Optical Nyquist Pulse Synthesizer Digital-to-Analog Converter

modeling, the optical behavior according to (4.6) is modeled by lossy waveguides and a charge-to-phase-shift-converter (q-to-$\Delta\phi$). As the name of the converter indicates, the phase shift $\Delta\phi$, but also the attenuation coefficient α, are depending on the charges q, stored in the ET-PS [6]:

$$\Delta\phi(q) = \pi \cdot \frac{q}{Q_\pi} \tag{4.12}$$

$$\Delta\alpha(q) = \ln(\sqrt{2}) \cdot \frac{q}{Q_{3dB} \cdot l} \tag{4.13}$$

with q being the excess charge in the ET-PS, Q_π being the excess charge required to cause a π phase shift, and Q_{3dB} being the excess charge required to cause a 3 dB-attenuation.

The excess charge q inside the ET-PS can be described by means of the forward transit time τ_{TT}, which models the time that the excess charge carriers need to leave the pn-junction [6, 9]:

$$q = \tau_{TT} \cdot i_D \tag{4.14}$$

It is also the cause of the diffusion capacitance c_d, as it represents the charges stored in the capacitance:

$$c_d = \frac{dq}{dv_D} = \tau_{TT} \cdot \frac{di_D}{dv_D} + i_D \cdot \frac{d\tau_{TT}}{dv_D} \approx \frac{\tau_{TT,on}}{r_d} \tag{4.15}$$

under the assumption of approximately constant forward transit time $\tau_{TT} \approx \tau_{TT,on}$ and $r_d = \frac{dv_D}{di_D}$ being the small-signal diode resistance [6].

With (4.6), (4.12) and (4.13) the ET-PS output E_{out} and the MZM transfer function A_{MZM} follow as:

$$E_{out}(t, l, v_{PS}) = E_0 \cdot e^{j\omega t} \cdot e^{-j\phi_0} \cdot e^{-l\alpha_0} \cdot e^{-j\pi \cdot \frac{q}{Q_\pi}} \cdot 2^{-\frac{q}{2 \cdot Q_{3dB}}} \tag{4.16}$$

$$A_{MZM} = 2^{-\frac{\bar{q}}{2 \cdot Q_{3dB}}} \cdot \frac{1}{4} \cdot |e^{-j(\frac{\pi}{2}\frac{\Delta q}{Q_\pi} + \frac{\Delta\phi_{th}}{2})} \cdot 2^{-\frac{\Delta q}{4 \cdot Q_{3dB}}} + e^{j(\frac{\pi}{2}\frac{\Delta q}{Q_\pi} + \frac{\Delta\phi_{th}}{2})} \cdot 2^{\frac{\Delta q}{4 \cdot Q_{3dB}}}|^2 \tag{4.17}$$

with $\bar{q} = (q_{PS,1} + q_{PS,2})/2$ and $\Delta q = q_{PS,1} - q_{PS,2}$.

The precise modeling of the ET-PS is important, because it allows application of an equalizer circuit to enhance the 3 dB-BW. One implementation of an equalizer circuit is shown in Fig. 4.14a. As Fig. 4.14b shows, the modulation efficiency decreases with increased equalization resistance R_{EQ}, which is affordable since the ET-PS exhibits very high efficiency. When adding the equalization capacitance C_{EQ} in parallel to R_{EQ}, the efficiency drop due to R_{EQ} decreases towards high frequency, resulting into peaking and BW extension [17]. Setting the RC equalizer to $R_{EQ} \cdot C_{EQ} = \tau_{TT}$, the current through C_{EQ} compensates the parasitic current through the diffusion capacitance c_d, which is resulting into a flat frequency response [6].

Using ET-PSs in conjunction with equalizer structures, allows similar characteristics as DT-PS MZMs at much smaller footprint. Therefore, ET-PSs enable much more complex photonic systems with multiple MZMs, such as the PONyDAC system in one EPIC.

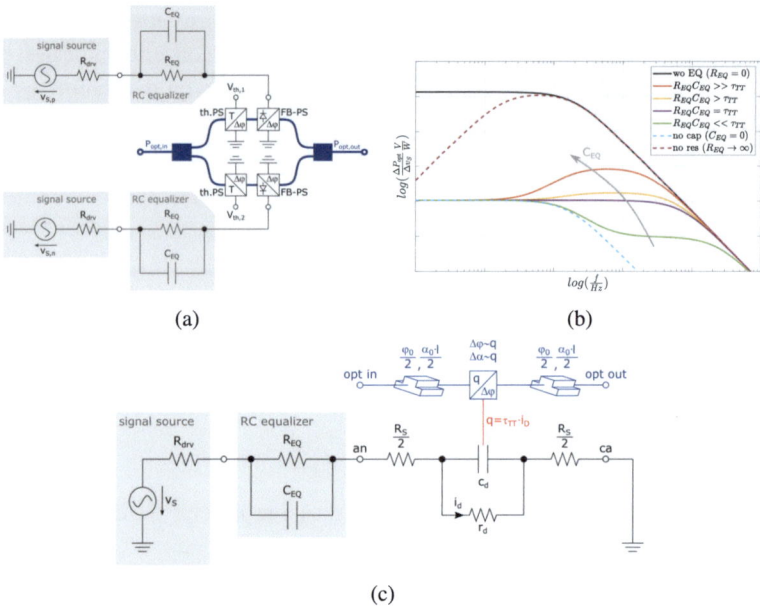

Fig. 4.14 ET-PS based MZM with passive RC equalizer **a** block diagram **b** frequency response for different RC sizings (substrate capacitance and contact inductance neglected) [6] **c** small-signal equivalent circuit

Measurements of a passively equalized MZM with 400 μm long ET-PSs, which are more than 15 times smaller than the DT-PS used in the MZM presented in Sect. 4.3.1, showed a 3 dB-BW of 17 GHz (presented in Fig. 4.15a). Simulations have revealed the possibility of enhancement up to 25 GHz, when improving the equalizer capacitance setting. As shown in Fig. 4.15b, data transmission experiments, have been conducted as well and show data transmission up to 50 Gb/s. The MZM was manufactured in a 250 nm BiCMOS technology which allows further design of area-efficient high-speed electro-optical transmitters with ET-PS MZMs and driver circuits, including electronic BW extension techniques, in one monolithically integrated chip.

4.4 Theoretical Investigations on the PONyDAC Performance

In order to assess the overall performance of the system and compare it with conventional, purely electronic DACs, the PONyDAC system was analyzed in terms of jitter, relative intensity noise (RIN), and nonlinearity [19]. Based on these results, estimates were made for achievable signal-to-noise-and-distortion ratios (SINADs) and effective number of bits (ENOB) (Fig. 4.16). The output's signal level, the transferred jitter from the clock signal, the transferred RIN caused by the optical source and distortion due to MZM non-linearity

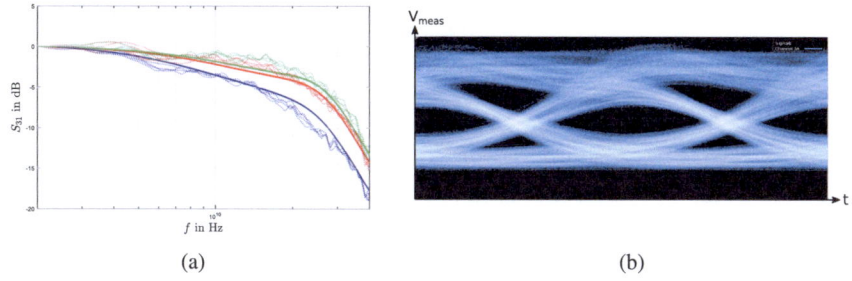

(a)   (b)

Fig. 4.15 ET-PS based MZM measurements **a** electro-optical S_{31}-parameter at different equalizer capacitances (dotted: measured, solid: simulated) **b** Eye diagram measured at 50 Gb/s data rate in NRZ modulation scheme (100 mV/div, 4 ps/div)

have been derived as [19]:

$$Signal(f) = P_{CW} H(f) G(V_S) \tag{4.18}$$

$$Jitter(f) = \frac{2}{3} P_{CW} (2\pi f)^2 \sigma_{t_j}^2 \tag{4.19}$$

$$RIN = \int_0^\infty S_{RIN}(f) H(f) \, df \quad \text{with} \quad S_{RIN} \approx \frac{\overline{\delta P_0^2(t)}}{B} \tag{4.20}$$

$$Distortion(f) = \sum_{\substack{k=-\infty \\ k \neq 0,1}}^{\infty} (\sin(\Phi) J_{2k-1}(\alpha))^2 + \sum_{\substack{k=-\infty \\ k \neq 0}}^{\infty} (\cos(\Phi) J_{2k}(\alpha))^2 \tag{4.21}$$

for an electrical input signal $v_i = V_S \cdot \sin(\omega_e t + \Phi_e)$ and optical MZM output $E_{out} = E_0 \cdot \cos(\pi \frac{v_i}{V_\pi})$, the normalized maximum optical output power P_{CW}, the BW $B = f_{max} - f_{min}$ of an applied optical bandpass filter with transfer function $H(f)$, $G(V_S)$ being a correction function for the optical peak power depending on the modulation amplitude V_S, the standard deviation of a random jitter process in the clock source $\sigma_{t_j} = \sqrt{2 \int_{f_1}^{f_2} \frac{1}{(2\pi f)^2} S_{PS}(f) df}$, the RIN's one-sided spectral density of the $S_{RIN}(f)$, the RIN's mean square power fluctuation $\overline{\delta P_0^2(t)}$, the Bessel function of n-th order J_n, and the normalized amplitude $\alpha = \pi \cdot \frac{V_S}{V_\pi}$.

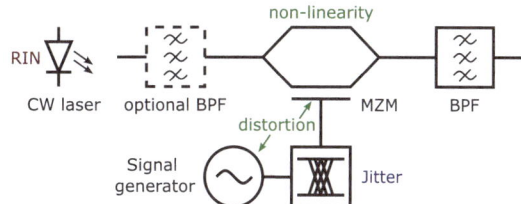

Fig. 4.16 Considered distortions and noise sources in the PONyDAC system [19]

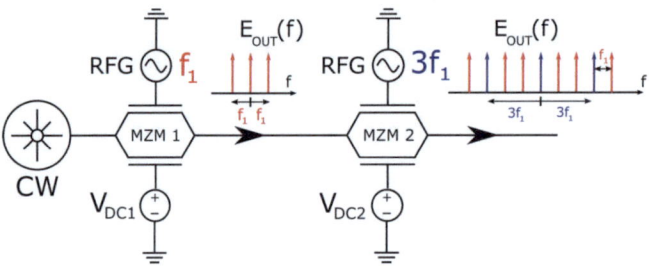

Fig. 4.17 Nine-line-comb NPS generation with cascaded MZMs [23]

Therefrom an estimation of the expected SINAD and ENOB was given with:

$$SINAD(f) = \frac{Signal(f)}{Jitter(f) + RIN + Distortion(f)} \quad (4.22)$$

$$ENOB(f) = \frac{10\,\text{dB}\log(SINAD(f)) - 1.77}{6.02} \quad (4.23)$$

predicting, that a PONyDAC system operated with an ultra-low phase noise signal generator [20], a commercially available continuous wave laser source with -165 dBc RIN [21], and a 1 GHz pre-filter enables more than 8 ENOB over more than 100 GHz BW [19]. Compared to the best electronic DACs, which are achieving 4.1 ENOB over 58.6 GHz BW [22], this is an enormous improvement in DAC quality.

4.5 Experimental Results

4.5.1 NPS Generation

Experiments have been carried out to explore the limits of DT-PS based MZMs described in Sect. 4.3.1. Figure 4.17 shows a setup, which was applied for generating three and nine line frequency combs. The frequency combs were measured using an optical spectrum analyzer (OSA) and the corresponding time-domain NPSs were recorded using an oscilloscope.

The first step involves generating a flat three-line frequency comb. This is achieved by driving an MZM at a frequency f_1 (MZM-1 in Fig. 4.17) and carefully adjusting its DC bias so that the three output spectral lines have equal power, as illustrated by the red spectrum in the inset of Fig. 4.17. In the experiment, the MZM was driven with a 20 GHz sinusoidal signal, and the DC bias was set appropriately. The resulting optical spectrum, measured using an OSA, is shown in Fig. 4.18a. This optical signal was then converted into the electrical domain using a high-speed photodiode, and the corresponding electrical waveform was captured with an oscilloscope. The measured signal forms NPSs with two zero crossings

4 Precise Optical Nyquist Pulse Synthesizer Digital-to-Analog Converter

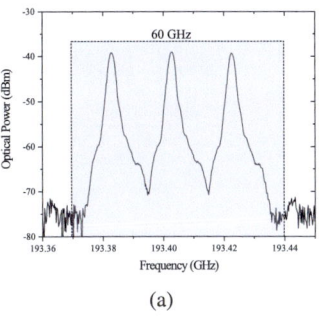
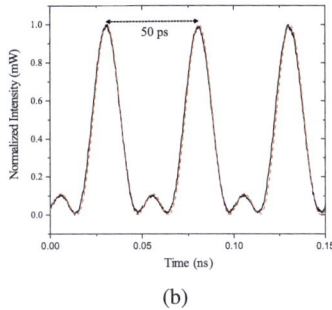

Fig. 4.18 Measured 60 GHz BW **a** three line frequency comb **b** NPS in time domain

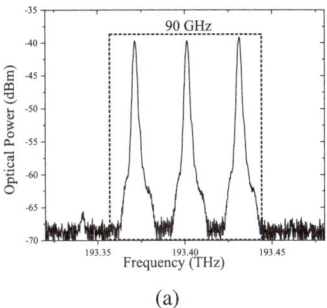
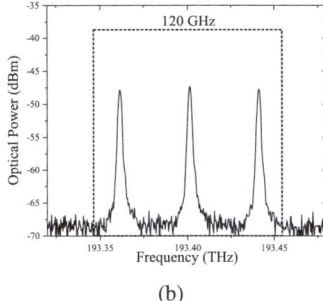

Fig. 4.19 Three line frequency comb with **a** 90 GHz BW and **b** 120 GHz BW

and a repetition period of 50 ps, as depicted in Fig. 4.18b. The experimentally generated pulse is also compared with a simulated ideal NPS, shown by the red dotted lines.

NPS generation is a critical building block for the PONyDAC concept, as discussed in detail in Sect. 1.2. A key advantage of this approach is the ability to easily vary the electrical driving signal, which directly impacts the BW of the resulting frequency comb and the repetition rate of the generated pulses. To demonstrate the performance of the proposed Si MZM, which has a limited 3 dB-BW of 17 GHz, the modulator was successfully driven with 30 GHz and 40 GHz sinusoidal signals. This resulted in the generation of three-line frequency combs with BWs of 90 GHz and 120 GHz, respectively. These results represent record BW outcome for a Si modulator and were made possible by the unique driving mechanism employed with DT-PS. Figure 4.19 shows three-line frequency combs with a BW of 90 GHz and 120 GHz, respectively. Due to the limited BW of the photodiode and oscilloscope these high BW pulses were not measured. However, the optical spectrum shows good signal to noise ratio and flatness of the spectral lines resulting in close to ideal NPSs as shown in Fig. 4.18b.

As described in Sect. 4.2, the NPS generation concept can be extrapolated to generate high-order frequency combs with the help of additional MZMs cascaded to the first modulator. With matching bias adjustment, two cascaded MZMs, which are modulated by sinusoidal

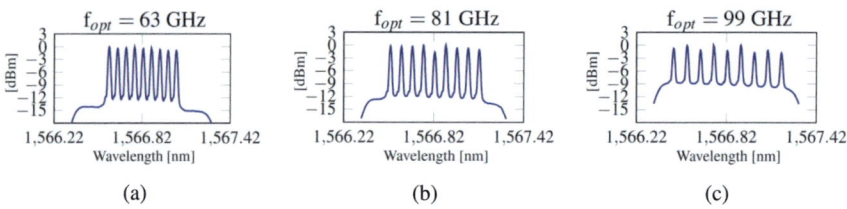

Fig. 4.20 Measured optical spectra [23] **a** $f_{opt} = 63$ GHz **b** $f_{opt} = 81$ GHz **c** $f_{opt} = 99$ GHz

signals with frequencies f_1 and $3 \cdot f_1$ (see Fig. 4.17), result into a nine-line frequency comb output spectrum and corresponding NPSs with eight zero crossings in time domain. As shown in Fig. 4.20, with careful adjustments of biasing voltages and electrical signal amplitudes, optical BWs of 63 GHz, 81 GHz, and 99 GHz have been achieved with repetition rates of 7 GHz, 9 GHz, and 11 GHz, respectively [23].

These results proof that the Si based EPIC implementations are also suitable for nine-channel PONyDAC systems and optical BWs of nearly 100 GHz can be reached.

4.5.2 PONyDAC Simulation and Experiment Results

The proposed PONyDAC concept was simulated using the OptiSystem software package. The simulation setup mirrored the conceptual design, where each channel included two modulators, as illustrated in Fig. 1.3.

The objective was to generate a 300 Gbaud 7-bit Pseudorandom Binary Sequence (PRBS) data signal using low-speed electronic and photonic components. A CW laser operating at a wavelength of 1550 nm and an optical power of 10 dBm served as the optical source. Its output was divided into three branches using a power splitter. In each branch, the first MZM was driven by a 100 GHz sinusoidal signal to generate sinc-shaped pulse sequences following the same approach previously demonstrated using the Si modulator. This resulted in a three-line frequency comb with 100 GHz spacing between spectral lines, corresponding to a 300 GHz NPS. The second MZM in each branch was used to modulate these pulses with predefined 100 Gbaud sinc-shaped data patterns. These data patterns were BW-limited to 50 GHz in the frequency domain to prevent spectral overlap when convolved with the three-line frequency comb. Accordingly, the first modulator required a BW of 100 GHz, while the second modulator needed a BW of 50 GHz.

To generate the targeted 300 Gbaud/s bit sequence, every third bit from the original sequence was selected to create predefined data patterns suitable for generation using arbitrary waveform generators (AWGs) operating at a 100 GS/s sampling rate. In the first branch, the predefined data sequence was multiplied with the sinc-shaped pulse sequence in the time domain. Similarly, in the second and third branches, the sinc pulses generated by the first modulators, which are having the same BW, were weighted with data patterns derived from the second and third bits of the original sequence, respectively. This approach effectively distributed the high-speed data across three parallel lower-speed channels.

After combining the three channels of weighted orthogonal pulses using an optical power combiner, the desired arbitrary waveform was achieved, tripling the sampling rate of the integrated electronics and photonics. For measuring the intensity and amplitude of the multiplexed signal, direct detection and coherent detection were employed, respectively. To compare with an ideal signal trace, the blue curve in Fig. 4.21 illustrates a 300 Gbaud/s PRBS-7 NRZ signal filtered with a 150 GHz low-pass rectangular filter, representing the target waveform. This requires an AWG with a high BW of 150 GHz and a sampling rate of 300 GS/s. The best commercially available electrical AWGs from Keysight (M8199A, M8199B) offer an overall output analog BW of 70 GHz. These Keysight AWGs and Keysight VXG (1 MHzâŁ"-110 GHz) can be used for parallelization with 100 GHz integrated modulators due to the requirement for low-BW components.

The performance of the PONyDAC with just 100 GHz electronics and photonics is shown by the red curve in Fig. 4.21. As evident, the PONyDAC closely approximates the ideal case. However, slight variations in waveform quality are observed due to the type of electrical filtering and NPS shaping used in the simulation layout. The corresponding eye diagrams for the generated signal is shown in Fig. 4.22.

For experimental validation, the PONyDAC concept setup illustrated in Fig. 4.4 was implemented using commercially available fiber-optic components, as the proposed EPIC was still undergoing fabrication. To ensure stable power levels during the measurement of high-BW signals, the original configuration, consisting of a single light source and a 1×3 optical combiner was modified by employing three separate laser sources, as described in [25].

Each branch of the system included an MZM driven by a 10 GHz signal, which produced a 30 GHz NPS, as previously discussed. These pulse sequences were time-aligned and multiplexed to form three orthogonal pulses. The combined output resulted in a flattened waveform, as depicted in Fig. 4.23a. The slight ripples observed in the otherwise flat wave-

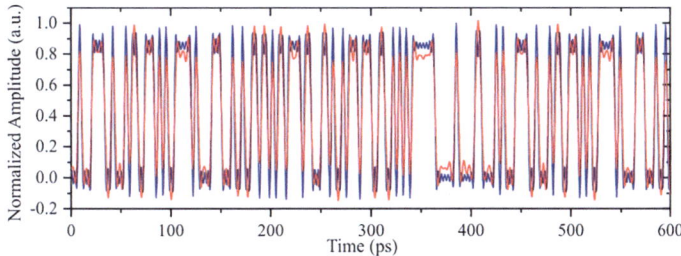

Fig. 4.21 300 Gbaud/s NRZ PRBS-7 pattern filtered with a 300 GHz rectangular filter (blue). The red curve shows the generated 300 Gbaud/s NRZ PRBS-7 data pattern using the PONyDAC with only 100 GHz electronics and photonics [24]

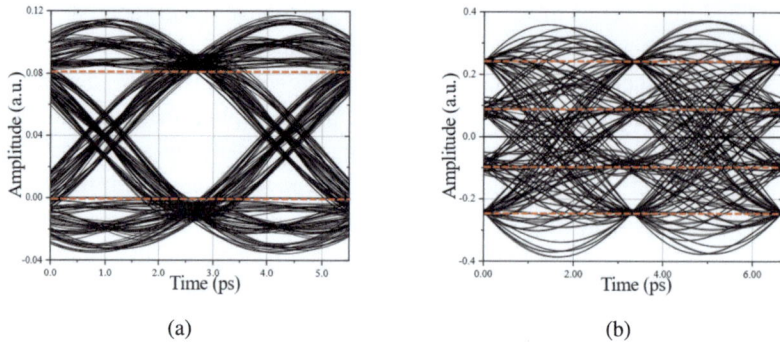

Fig. 4.22 Eye diagram for the generated **a** two level and **b** four level 300 Gbaud/s signal generated by using the three channel PONyDAC [24]

form are attributed to minor power mismatches among the branches. Such inconsistencies are expected to be eliminated in a fully integrated Si photonics prototype.

To synthesize a sinusoidal waveform, each 10 GHz pulse sequence in the three branches was weighted using a 10 GS/s binary sequence alternating between logical 1 and 0. The timing was adjusted so that the pulses aligned with the center of the data sequences, which are three times longer than the pulse duration. Superimposing these weighted sequences resulted in a 15 GHz sinusoidal waveform, shown in Fig. 4.23b. This demonstrates that using modulators with a 10 GHz RF BW, it is possible to generate a waveform with a frequency 1.5 times higher than the modulator BW. Furthermore, by leveraging 10 GS/s electronics across three channels, an effective sampling rate of 30 GS/s was achieved.

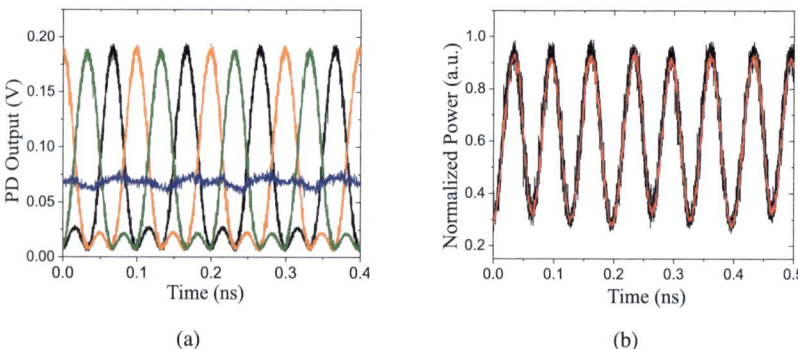

Fig. 4.23 The blue line illustrates the combined output of time-shifted orthogonal pulses from three channels, as depicted in (**a**). When these pulses are appropriately weighted and superimposed, they form a 15 GHz sinusoidal waveform, shown in black in (**b**). This result is compared against the theoretical sum of the weighted pulses across all three channels, shown in red in (**b**). Consequently, with three branches and modulators driven at 20 GHz, the system achieves an effective sampling rate of 30 GS/s [25]

4 Precise Optical Nyquist Pulse Synthesizer Digital-to-Analog Converter

The cascaded DT-PS based MZMs integrated within the EPIC platform were employed to emulate the complete PONyDAC architecture. In the first MZM, the optical carrier was modulated using a single-tone 10 GHz signal generated by a signal generator. By carefully adjusting the bias through temperature tuning, the optical carrier was suppressed, resulting in an output from MZM-1 that formed a flat frequency comb with three tones. The second on-chip MZM was then used to apply data weighting to these NPSs using a random bit sequence at a sampling rate of 10 GS/s, generated by an electronic AWG.

To emulate the second and third branches of the PONyDAC architecture, the phase of the signal generator was shifted by 120 and 240 degrees, respectively. The outputs from all three branches were mathematically combined to generate the waveforms shown in Fig. 4.24. A 15 GHz sinusoidal waveform was produced, as illustrated in Fig. 4.24a. The individual outputs from the three branches are displayed in black, green, and orange, while their summed result is shown in blue. The corresponding electrical spectrum is presented in Fig. 4.24b, exhibiting a signal-to-noise ratio (SNR) of 32.7 dB. Additionally, a sawtooth waveform was generated by modifying the amplitude of the data signal used to weight the

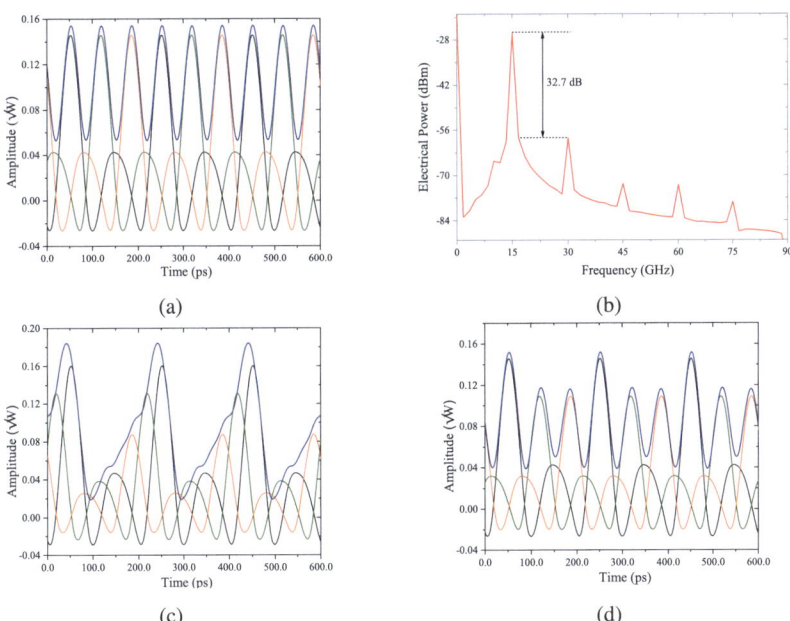

Fig. 4.24 The individual amplitudes of the three branches are shown in black, green, and orange, while their combined output is depicted in blue. The superposition of the three channels results in a 15 GHz sinusoidal waveform (**a**), with its corresponding electrical spectrum shown in (**b**). By adjusting the amplitude of the weighting signals, additional arbitrary waveforms were generated, as illustrated in (**c**) and (**d**) [26]

pulses in MZM-2, as shown in Fig. 4.24c. Due to the use of only three branches, the resulting ramp waveform does not exhibit a perfectly linear slope.

A partial implementation of the proposed PONyDAC concept based on the time-interleaved superposition of weighted sinc-pulse sequences has been demonstrated using two modulators integrated on an EPIC. Through mathematical post-processing of each channel, the system successfully generates arbitrary waveforms at an effective sampling rate three times higher than the modulation rate. To achieve even higher sampling rates, either faster modulators would have to be used or the number of channels would have to be increased. However, the latter would require either large chip areas with the associated costs, or miniaturized modulators, such as the forward-biased modulators investigated and presented in Sect. 4.3.2. Considering available high-speed integrated modulators with BWs beyond 100 GHz [27, 28] and appropriate electronic drivers, this architecture has the potential to achieve sampling rates of up to 300 GS/s. However, the driving limitation is again defined by the electronic signal and waveform generators. A fully integrated solution would offer low timing jitter and enhanced tunability in the electrical domain, making it a promising candidate to overcome the BW limitations of current state-of-the-art communication systems.

4.6 Conclusion

A new concept of a photonic-electronic DAC has been presented and investigated. Two possible topologies have been found and compared to each other. The topology of a PONyDAC with multiple pulse synthesizers allows higher flexibility at the costs of an increased amount of MZMs required to realize the system, compared to PONyDACs with single pulse synthesizers.

Moreover, two MZM types have been designed, investigated, and applied in EPIC technology. Respecting the size of the PONyDAC system, ET-PS based MZMs fits well to the overall system. Therefore, ET-PSs have been modeled and equalizer circuits have been applied to the PS, increasing the BW and raise to DT-PS performance levels at very much decreased size.

Three and nine-line NPSs have been generated in EPIC technology, which form the bases for the PONyDAC system and enable three and nine-fold sampling rates respectively.

Theoretical investigations and experimental results both prove the PONyDAC functionality. A PONyDAC demonstrator was built and used to generate various signal shapes with total sampling rates corresponding to a multiple of the electronics used. It is shown that with commercially available laser sources and filters, more than 8 ENOB over more than 100 GHz BW can be achieved, which is far better than state-of-the-art electronic DACs.

Therefore, the PONyDAC is compatible with silicon photonics platforms, shows very high applicability, and enables significantly faster DACs than have ever been realized in purely electronic technologies.

References

1. Shannon CE (1948) A mathematical theory of communication. Bell Syst Tech J 27(3):379–423
2. Soto MA, Alem M, Shoaie MA, Vedadi A, Brès C-S, Thénevaz L, Schneider T (2013) Optical sinc-shaped Nyquist pulses of exceptional quality. Nat Commun 20(1):1–11 (2013)
3. Misra A, Kress C, Singh K, Preußler S, Scheytt JC, Schneider T (2019) Integrated source-free all-optical sampling with a sampling rate of up to three times the RF bandwidth of silicon photonic MZM. Opt Exp 27(21):29972–29984
4. Misra A, Kress C, Singh K, Meier J, Schwabe T, Preussler S, Scheytt JC, Schneider T (2022) Reconfigurable and real-time high-bandwidth nyquist signal detection with low-bandwidth in silicon photonics. Opt Exp 30(8):13776–13789
5. Soref R, Bennett B (1987) Electrooptical effects in silicon. IEEE J Quantum Electron 23(1):123–129
6. Schwabe T, Kress C, Kruse S, Weizel M, Rhee H, Scheytt JC (2024) Forward-biased silicon phase shifter modelling for electronic-photonic co-simulation and validation in a 250 nm epic bicmos technology. J Lightwave Tech 1–16
7. Rahim A, Hermans A, Wohlfeil B, Petousi D, Kuyken B, Van Thourhout D, Baets RG (2021) Taking silicon photonics modulators to a higher performance level: state-of-the-art and a review of new technologies. Adv Photon 3(2):024003
8. Kress C, Schwabe T, Rhee H, Scheytt JC (2024) Compact, high-speed mach-zehnder modulator with on-chip linear drivers in photonic bicmos technology. IEEE Access 12:64561–64570
9. Reisch M (2007) Halbleiter-Bauelemente, chapter 2.3–2.5, 2nd ed. Springer-Lehrbuch. Springer, Berlin, Heidelberg, pp 73–85
10. Chung H, Chang WSC, Adler EL (1991) Modeling and optimization of traveling-wave linbo/sub 3/interferometric modulators. IEEE J Quantum Electron 27(3):608–617
11. Denoyer G, Cole C, Santipo A, Russo R, Robinson C, Li L, Zhou Y, Chen JA, Park B, Boeuf F, Crémer S, Vulliet N (2015) Hybrid silicon photonic circuits and transceiver for 50 gb/s nrz transmission over single-mode fiber. J Lightwave Technol 33(6):1247–1254
12. Kress C, Schwabe T, Rhee H, Kerman S, Scheytt CJ (2022) Broadband mach-zehnder modulator with linear driver in electronic-photonic co-integrated platform. In: Optica advanced photonics congress 2022, page IM4C.1. Optica Publishing Group
13. Knoll D, Lischke S, Awny A, Kroh M, Krune E, Mai C, Peczek A, Petousi D, Simon S, Voigt K, Winzer G, Barth R, Zimmermann L (2015) High-performance bicmos si photonics platform. In: 2015 IEEE bipolar/BiCMOS circuits and technology meeting—BCTM, pp 88–96
14. Kress C, Singh K, Schwabe T, Preußler S, Schneider T, Scheytt JC (2021) High modulation efficiency segmented mach-zehnder modulator monolithically integrated with linear driver in 0.25 μm bicmos technology. In: OSA advanced photonics congress 2021, page IW1B.1. Optica Publishing Group
15. Jaeger R, Blalock T (2010) Microelectronic circuit design, chapter 17.6.1 The miller effect and 17.10.3 high-frequency response of the cascode amplifier, 4th ed. McGraw-Hill Education, pp 1159–1162, 1184–1185
16. Gray P, Hurst P, Lewis S, Meyer R (2009) Analysis and design of analog integrated circuits, chapter 3.36 common-collector configuration and 7.2.3.1 frequency response of the emitter follower, 5th ed. Wiley, pp 191–194, 505–511

17. Baba T, Akiyama S, Imai M, Usuki T, Usuki T (2015) 25-gb/s broadband silicon modulator with 0.31vcm vpil based on forward-biased pin diodes embedded with passive equalizer. Opt Exp 23(26):32950–32960
18. Sobu Y, Huang G, Mori T, Tsunoda Y, Yamamoto T, Tanaka S, Hoshida T (2022) Highly power-efficient (2 pj/bit), 128gbps 16qam signal generation of coherent optical dac transmitter using 28-nm cmos driver and all-silicon segmented modulator. In: 2022 optical fiber communications conference and exhibition (OFC), pp 1–3
19. Kress C, Bahmanian M, Schwabe T, Scheytt JC (2021) Analysis of the effects of jitter, relative intensity noise, and nonlinearity on a photonic digital-to-analog converter based on optical nyquist pulse synthesis. Opt Exp 29(15):23671–23681
20. Bahmanian M, Fard S, Koppelmann B, Christoph Scheytt J (2020) Wide-band frequency synthesizer with ultra-low phase noise using an optical clock source. In: 2020 IEEE/MTT-S international microwave symposium (IMS), pp 1283–1286
21. More Photonics. Apic cwl-100-1550-165 ultra-low rin dfb laser (2024)
22. Collisi M, Möller M (2020) A 120 gs/s 2:1 analog multiplexer with high linearity in sige-bicmos technology. In: 2020 IEEE BiCMOS and compound semiconductor integrated circuits and technology symposium (BCICTS), pp 1–4
23. Kress C, Schwabe T, Silberhorn C, Christoph Scheytt J (2023) Generation of 100 ghz periodic nyquist pulses using cascaded mach-zehnder modulators in silicon electronic-photonic platform. In: CLEO 2023, page SF1P.6. Optica Publishing Group
24. Singh K, Meier J, Preußler S, Kress C, Scheytt JC, Schneider T (2021) Optical prbs generation with threefold bandwidth of the employed electronics and photonics. In: OSA advanced photonics congress 2021, page SpTu4D.6. Optica Publishing Group
25. Singh K, Meier J, Misra A, Preußler S, Scheytt JC, Schneider T (2020) Photonic arbitrary waveform generation with three times the sampling rate of the modulator bandwidth. IEEE Photon Technol Lett 32(24):1544–1547
26. Singh K, Meier J, Kress C, Misra A, Schwabe T, Preußler S, Scheytt JC, Schneider T (2022) Emulation of integrated high-bandwidth photonic awg using low-speed electronics. In: Next-Generation optical communication: components, sub-systems, and systems XI, vol 12028, pp 42–48
27. Valdez F, Mere V, Wang X, Mookherjea S (2023) Integrated o- and c-band silicon-lithium niobate mach-zehnder modulators with 100 ghz bandwidth, low voltage, and low loss. Opt Exp 31(4):5273–5289
28. Yue H, Jianbin F, Zhang H, Xiong B, Pan S, Chu T (2025) Silicon modulator exceeding 110 ghz using tunable time-frequency equalization. Optica 12(2):203–215

Open Access This chapter is licensed under the terms of the Creative Commons Attribution 4.0 International License (http://creativecommons.org/licenses/by/4.0/), which permits use, sharing, adaptation, distribution and reproduction in any medium or format, as long as you give appropriate credit to the original author(s) and the source, provide a link to the Creative Commons license and indicate if changes were made.

The images or other third party material in this chapter are included in the chapter's Creative Commons license, unless indicated otherwise in a credit line to the material. If material is not included in the chapter's Creative Commons license and your intended use is not permitted by statutory regulation or exceeds the permitted use, you will need to obtain permission directly from the copyright holder.

Integrated Low Jitter Mode-Locked Lasers and Pulse/Spectrum Shapers

Jeremy Witzens, Milan Sinobad, Tengizi Abramishvili, Pascal Gehrmann, Andrea Zazzi, Mike Külkens, Jan Lorenzen, Alvaro Moscoso Mártir, Neetesh Singh and Franz X. Kärtner

Abstract

This chapter discusses progress made towards the chip-scale integration of ultra-low jitter mode-locked lasers using rare-earth gain media with CMOS compatible fabrication. Such lasers generate ultra-stable pulse trains and optical frequency combs that are instrumental for signal processing systems such as optically assisted analog-to-digital converters. We also review work made towards the co-integration of optical filters for the interleaving of pulses or the upconversion of the comb's free spectral range, as a further step towards a complete system integration.

J. Witzens (✉) · T. Abramishvili · A. Moscoso Mártir
Institute of Integrated Photonics, RWTH Aachen University, Aachen, Germany
e-mail: jwitzens@iph.rwth-aachen.de

M. Sinobad · M. Külkens · J. Lorenzen · F. X. Kärtner
Physics Department, University of Hamburg, Hamburg, Germany

M. Sinobad · P. Gehrmann · M. Külkens · J. Lorenzen · N. Singh · F. X. Kärtner
Center for Free-Electron Laser Science, Deutsches Elektronen-Synchrotron, Hamburg, Germany

P. Gehrmann
Institute of High-Frequency Technology, Hamburg University of Technology, Hamburg, Germany

A. Zazzi
University of California at Berkley, Berkley, CA, USA

M. Külkens
Cycle GmbH, Hamburg, Germany

Max Planck Institute for the Structure and Dynamics of Matter, Hamburg, Germany

© The Authors(s) 2026
J. C. Scheytt et al. (eds.), *Electronic-Photonic Integrated Systems for Ultrafast Signal Processing*, https://doi.org/10.1007/978-3-032-08340-1_5

Acronyms

ADC	Analog-to-Digital Converters
ACG	Apodized Chirped Bragg Grating
ASE	Amplified Spontaneous Emission
CMOS	Complementary Metal-Oxide-Semiconductor
CROW	Coupled Resonator Optical Waveguide
DCS	Directional Coupler Splitter
DSP	Digital Signal Processing
DUV	Deep Ultraviolet
EDFA	Erbium-Doped Fiber Amplifier
ENOB	Effective Number of Bits
FSR	Free Spectral Range
FWHM	Full Width at Half Maximum
GD	Group Delay
GDD	Group Delay Dispersion
IL	Insertion Loss
LPCVD	Low Pressure Chemical Vapor Deposition
LiDAR	Light Detection And Ranging
LMR	Large Mode Area
MLL	Mode-locked Lasers
MIMO	Multiple-Input Multiple-Output
NLI-SA	Nonlinear Michelson Interferometer-based Saturable Absorber
NF	Noise Figure
OADM	Optical Add-Drop Multiplexer
OFC	Optical Frequency Comb
ORR	Optical Ring Resonator
OSA	Optical Spectrum Analyzer
PIC	Photonic Integrated Circuit
Q-factor	Quality Factor
RF	Radio Frequency
RMS	Root-Mean-Square
SNR	Signal-to-Noise Ratio
SiN	Silicon Nitride
TE	Transverse Electric
TI	Time-Interleaved
TM	Transverse Magnetic
WDM	Wavelength-Division Multiplexing

5.1 Introduction

Optical pulse trains with ultra-low jitter and phase noise are essential for many high-performance signal processing applications. Examples include precision light detection and ranging (LiDAR) for autonomous vehicles, optical clocks for satellite-based navigation, optical frequency synthesis for scientific metrology and high-speed, optically enabled analog-to-digital converters (ADCs). One of the most effective ways to generate such signals consists in using mode-locked lasers (MLLs), which produce periodic trains of ultrashort pulses with repetition rates ranging from 100 MHz to several GHz. In the spectral domain, these pulse trains correspond to evenly spaced, mutually coherent lines that form a frequency comb. Upon photodetection, the generated microwave comb corresponds to a set of equally spaced beat-tones that can be used as high frequency reference signals. Optically pumped MLLs based on rare-earth-doped amplifiers are particularly well-suited for applications requiring low jitter and phase noise, since they support femtosecond pulses and provide high power from lasers with large mode areas and extended cavity lengths. As explained in Sect. 5.2, this suppresses the influence of noise sources on the timing jitter of the pulse train and improves timing stability.

In systems that rely on optical pulse trains with ultra-low phase noise—such as those employing MLLs for optical sampling, clock distribution, or frequency synthesis—the timing jitter of the pulse train is a critical performance metric. It refers to variability in the arrival time of individual pulses relative to an ideal periodic pulse train. When such a pulse train is used to generate a radio-frequency (RF) signal through photodetection, timing jitter manifests as the synthesized signal's phase noise, which directly impacts the performance of high-speed ADCs whose achievable resolution is fundamentally limited by the aperture jitter inherited from the sampling clock [1].

MLLs are a particularly well-suited technology for this application because they can have much less timing jitter than conventional microwave oscillators that operate at similar power levels. This performance is the result of basic physical properties of ultra-short optical pulses, which concentrate a large amount of energy into an extremely short time interval. This combination of high pulse energy and short pulse duration makes the pulse timing less susceptible to noise. Furthermore, rare-earth doped gain media reduce the coupling between gain and refractive index fluctuations and, therefore, pulse timing. The quantum limit of timing jitter increases proportionally to the pulse duration and decreases with the square root of the pulse energy [2]. This favors designs that maximize the energy stored in the gain media and provide a large gain bandwidth to support ultrashort pulses. Theoretical and experimental studies have shown that table-top femtosecond MLLs can produce optical pulse trains with timing jitter of a few attoseconds, integrated over offset frequencies above one kHz [3]. Due to the scaling of jitter with intracavity pulse energy and pulse duration, one can expect that on-chip femtosecond lasers can still achieve sub-fs-level jitter.

The lowest jitter femtosecond sources demonstrated to date are based on either free-space solid-state or fiber-based lasers. These systems are bulky, sensitive to alignment, and difficult to scale. Therefore, these platforms are difficult to use for emerging applications that demand compact, robust, and thermally stable devices, such as mobile frequency metrology, integrated radar, and chip-scale microwave sources and digitizers. This has motivated intensive efforts to implement low-jitter MLLs at the chip scale, especially in CMOS-compatible platforms for cost-reduction and for co-integration with electronics at a later stage. Integration provides several advantages beyond footprint reduction. It enables the integration of electronic and photonic subsystems, which is essential for realizing photonically assisted ADCs. It also allows for monolithic co-integration of stabilization loops for scalable manufacturing. Specific architectural choices—such as the use of large-mode-area waveguide amplifiers—support higher pulse energies while limiting nonlinear effects and provide a path towards increasing the performance of integrated chip-scale devices. The integration of rare-earth gain media in CMOS chips is another essential aspect, since they provide both the wide gain spectrum and the low noise figure required for these applications. While progress has been made towards reducing the jitter of semiconductor MLLs, by combining the semiconductor gain medium with low loss intra-cavity [4] or extra-cavity [5] optical delay lines at the chip scale, the high nonlinearity and limited gain spectrum of these devices fundamentally limits peak power, i.e. pulse energy divided by pulse duration, and, therefore, severally limits the performance with regard to jitter.

In this chapter, we present progress towards the design and fabrication of an integrated low-jitter MLL using rare-earth gain media, as well as its integration with on-chip pulse and spectral shapers. First, we review the origin and scaling of timing jitter in ultra-short-pulse MLLs (Sect. 5.2), architectural strategies for jitter suppression in chip-scale devices (Sect. 5.3), and recent experimental progress towards integrated devices with performance approaching that of fiber-based systems (Sect. 5.4). We then give an example of an application in the form of a time-interleaved, photonically-assisted, high-speed ADC (Sect. 5.5) and focus, in particular, on the implementation of a suitable pulse shaper that can be integrated in the same photonic integrated circuit (PIC) technology as the laser. For applications that require a free spectral range (FSR) in the frequency domain that is wider than what can be achieved at the chip scale [6–8], we implement an on-chip filter in a CMOS compatible silicon nitride technology that can also be cascaded with the MLL on the same PIC. The concept of the filter is described in Sect. 5.6, with details of its implementation in Sect. 5.6.2 and first experimental results in Sect. 5.6.3. We summarize the chapter in Sect. 5.7.

5.2 Timing Jitter in Mode-Locked Lasers: Fundamentals, Scaling Laws, and Comparison with Electronic Oscillators

The origin of timing jitter and the physical mechanisms that govern timing fluctuations in MLLs can be best understood by comparing them to noise processes in conventional microwave oscillators. In this section, we examine how phase noise arises in microwave systems and how timing jitter builds up in MLLs, highlighting similarities and differences in noise scaling between the two. We introduce key physical parameters that determine jitter performance, such as pulse duration, pulse energy, and the amount of noise added to the intra-cavity field per round trip. Figure 5.1 illustrates the main differences by comparing the waveform of a sinusoidal microwave oscillator with that of an ultra-short pulse train from an MLL, in which temporal energy confinement plays a key role. We further quantify this comparison with experimental data reported in the literature.

A microwave oscillator consists of a resonator and an amplifier in a feedback loop configuration. The resonator supports an oscillating electromagnetic mode at a specific frequency with stored energy W_{mode} in steady state oscillation. As the oscillator generates a microwave signal, part of this energy is emitted through the output port, while additional losses occur internally due to dissipation in the cavity. The amplifier must compensate for both the out-coupled power and the intrinsic cavity losses to maintain steady-state oscillation. The loaded quality factor, Q, is defined as the ratio of the energy stored in the resonator to the energy lost during one radian of the oscillation cycle. The corresponding cavity decay time, $\tau_{\mathrm{cav}} = QT_0/2\pi$, where T_0 is the oscillation period, characterizes how quickly the stored energy would decay in the absence of amplification. We use this decay time not only to describe how energy leaves the cavity, but also to calculate the amount of thermal noise that must be injected to maintain equilibrium, since decaying thermal noise is replenished by the environment. The coupling of the cavity to loss reservoirs leads to thermalization of the

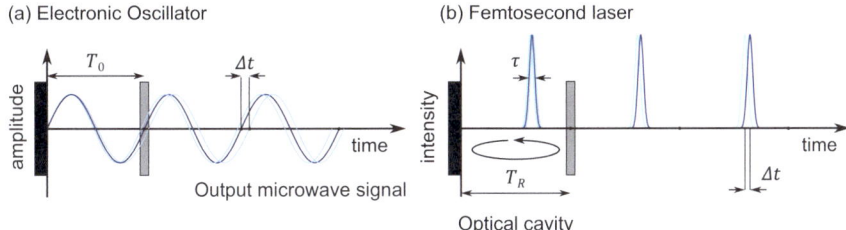

Fig. 5.1 **a** In an electronic oscillator, timing is defined by the zero crossings of a microwave signal with period T_0. The underlying reference oscillator (dark blue curve) indicates a noise-free signal, while the light blue curves represent a signal that suffers from phase advance or phase retardation that builds up over time up due to noise. **b** In a femtosecond laser, ultrashort pulses of duration τ are emitted at regular intervals T_R. The short temporal extent of each pulse limits the time window during which noise can directly perturb the pulse timing, effectively reducing the impact of the noise per cavity round-trip

field in the passive resonator. At equilibrium, the average thermal energy stored in the cavity mode is $k_B T$, where k_B is the Boltzmann constant and T is the cavity temperature in Kelvin [9]. This results in a continuous thermal power loss of $k_B T / \tau_{cav}$, which must be balanced by an equal thermal noise power entering the cavity from the environment. In an oscillator with active elements, this thermal noise influx is at least doubled by the noise of the feedback amplifier that compensates for the cavity loss. This additional noise contribution that arises from non-ideal amplifier behavior is described by a dimensionless excess noise factor Θ.

In such systems, the rate at which the mean square timing jitter grows is determined by the amount of thermal and amplifier noise power injected into the oscillating mode relative to the energy stored in it. The inverse dependence on W_{mode} arises from the fact that thermal noise perturbs the resonator field with a fixed power, so that a larger stored signal energy reduces the relative impact of these fluctuations on the signal phase. Therefore, greater stored signal energy leads to improved timing stability. Both the stored energy W_{mode} and the cavity decay time τ_{cav} are proportional to the cavity quality factor Q. Consequently, timing jitter decreases as Q^2 increases. Longer oscillation periods T_0 lead to a larger jitter growth rate, since timing jitter is derived from phase fluctuations divided by ω_0, with a variance thus scaling as $1/\omega_0^2 = T_0^2/(4\pi^2)$. Taking into account that only half of the noise power contributes to phase noise [9], the resulting expression for the rate at which the timing jitter Δt_{RF} grows is given by [10]

$$\frac{d}{dt} \langle \Delta t_{RF}^2 \rangle = \frac{T_0^2}{4\pi^2} \cdot \frac{\Theta}{W_{mode}} \cdot \frac{k_B T}{2\tau_{cav}} . \qquad (5.1)$$

An MLL delivers a train of ultrashort optical pulses with duration τ at a repetition rate of $1/T_R$, where T_R is the cavity round-trip time. The intracavity pulse energy is denoted by W_{pulse}. Similarly to the case of microwave oscillators, the optical cavity experiences losses from both internal processes and output coupling, which must be compensated by the gain. These losses determine the optical cavity decay time τ_{cav}, defined as the rate at which energy exits the cavity. The main source of noise in this system is the spontaneous emission introduced by the optical amplifier, which compensates for both internal losses and output coupling. The injected noise power is proportional to $h\nu_s/\tau_{cav}$, where ν_s is the optical carrier frequency and h is Planck's constant. Consequently, the increase in pulse position uncertainty per round-trip is proportional to the amplified spontaneous emission (ASE) noise power $\Theta h\nu_s/\tau_{cav}$, and inversely proportional to the intracavity pulse energy. This is the quantum analogue of the thermal noise limit in microwave oscillators, with $k_B T$ replaced by the photon energy $h\nu_s$, and Θ again representing an excess noise factor that captures non-ideal amplifier noise.

Importantly, H. A. Haus and A. Mecozzi derived that the variance of the timing jitter is not only influenced by the amplifier and cavity parameters, but also scales with the square of the pulse duration τ^2 [2]. This dependence reflects the fact that shorter pulses interact with the noise for a shorter duration, thereby reducing their timing uncertainty. This quadratic dependence on τ is a key distinction from the case of microwave oscillators, in which the variance of the timing jitter grows instead as T_0^2. Without going into the details of the different regimes of mode locking, see for example [3], the scaling of timing jitter in MLLs generally scales as

$$\frac{d}{dt}\langle \Delta t_{\text{MLL}}^2 \rangle = \frac{\pi^2}{6}\tau^2 \cdot \frac{1}{W_{\text{pulse}}} \cdot \frac{h\nu_s}{\tau_{\text{cav}}} \quad (5.2)$$

where τ is taken as the full width at half maximum (FWHM) of the pulse divided by 1.76.

A comparison of Eqs. (5.1) and (5.2) reveals two key differences. The first concerns the fundamental noise level. Assuming similar stored energies and cavity decay times, the noise power in optical systems is higher than in microwave systems because of the ratio $h\nu_s/k_B T \approx 32$ at typical operating points ($\nu_s \approx 200$ THz, $T \approx 300$ K). Under typical conditions, quantum noise at optical frequencies is thus higher than thermal noise at microwave frequencies. However, the dominant factor determining the jitter performance of MLL pulse trains compared to that of microwave oscillators arises from the difference in time scales. The timing jitter in MLLs scales with the square of the pulse duration, whereas in microwave oscillators it scales with the square of the oscillation period. For example, for a 10 GHz microwave oscillator with $T_0 = 100$ ps and a femtosecond laser with a pulse width of 100 fs, the ratio $T_0^2/\tau^2 = 10^6$, or 60 dB. This dramatic difference in scaling explains why MLLs can achieve timing jitter levels that are orders of magnitude lower than even the best microwave oscillators, despite operating with higher intrinsic noise power. Taken together, the higher noise energy in optical systems is overwhelmed by the far shorter pulse duration possible at optical frequencies. Consequently, the timing jitter noise floor of MLLs is often 40 dB or more lower than that of typical microwave sources, with similar energy storage capabilities and cavity decay times.

5.2.1 Timing Jitter and Its Impact on ADC Performance

We have reviewed the origin of timing jitter and explained why femtosecond MLLs outperform electronic oscillators in terms of timing stability. Here we describe how this advantage improves performance in high-speed systems, especially in ADCs, in which timing jitter affects the achievable resolution. We quantify the relation between timing jitter and the effective number of bits (ENOB), and show the importance for high-resolution sampling of high-frequency signals.

A fundamental limitation in high-speed analog-to-digital conversion is the aperture jitter or the timing uncertainty in the sampling. This jitter arises from the phase noise/timing jitter of the sampling clock, typically derived from an electronic RF oscillator on chip. This limitation can be quantified using a relation relating the ENOB to the root-mean-square (RMS) timing jitter Δt and the input signal frequency f_{in} [11]:

$$\text{ENOB} = \frac{-20\log_{10}(2\pi f_{\text{in}} \Delta t) - 1.76}{6.02} \quad (5.3)$$

The logarithmic term represents the signal-to-noise ratio (SNR) degradation due to aperture jitter, which is plugged in the standard SNR-to-ENOB conversion formula. This expression shows that achieving high resolution (ENOB) at high signal frequencies requires extremely low sampling jitter.

A survey of electronic ADC performance [11], the widely referenced Walden plot, has been updated with a comprehensive review of more recent state-of-the-art systems and is shown in Fig. 5.2. Light blue circles correspond to the initial survey of electronic ADCs, dark blue circles are more recently reported electronic ADCs, and orange circles are photonic ADCs. The best electronic ADCs have achieved jitter levels slightly below 100 fs. Further jitter reduction in electronic architectures becomes increasingly challenging, especially as target frequencies move into the multi-GHz range. In contrast, photonic ADCs, marked in orange, have demonstrated some of the best jitter performance at high operating frequencies exceeding 35 GHz. However, these devices are often still implemented with discretete components and fiber based systems of limited scalability, thus sacrificing resolution for speed. Our goal is to close this gap with integrated MLLs capable of sub-10-fs timing jitter combined with photonic ADCs at the chip scale, thus combining high-speed operation with high-resolution sampling.

5.3 Architecture and Design of Integrated Mode-Locked Lasers

Attempts to integrate MLLs with rare-earth gain media directly onto silicon chips began in 2009, when the first successful demonstration of on-chip mode locking was achieved using doped silica glass waveguides for amplification [17]. In that system, the saturable absorber was not integrated into the chip and silica glass waveguides were used to guide light. A decade later, in 2019, a fully integrated silicon-based architecture was attempted, in which silicon nitride waveguides were used to guide light and implement an artificial on-chip saturable absorber, although amplification still took place in doped glass waveguides [18]. While this device operated in the Q-switched mode-locking regime, these promising results marked a further step forward towards stable mode locking directly on a silicon chip. In our most recent implementation, we used low-loss silicon nitride waveguides in combination with a hybrid gain section based on silicon nitride and a sputter-deposited oxide gain film [19]. In the absence of proper dispersion compensation, the device was exhibiting stable Q-switched behavior [20] with high output pulse energies of 150 nJ at a 1 MHz repetition rate. This architecture provides a foundation for the development of low noise integrated MLLs with reduced cavity loss and high average power, that should exhibit mode locking once dispersion control is implemented, for example by using ultra-broadband, dispersion-compensating apodized chirped Bragg gratings (ACGs). This architecture is illustrated in Fig. 5.3. In the following sections we discuss its key components in more detail.

5 Integrated Low Jitter Mode-Locked Lasers and Pulse/Spectrum Shapers

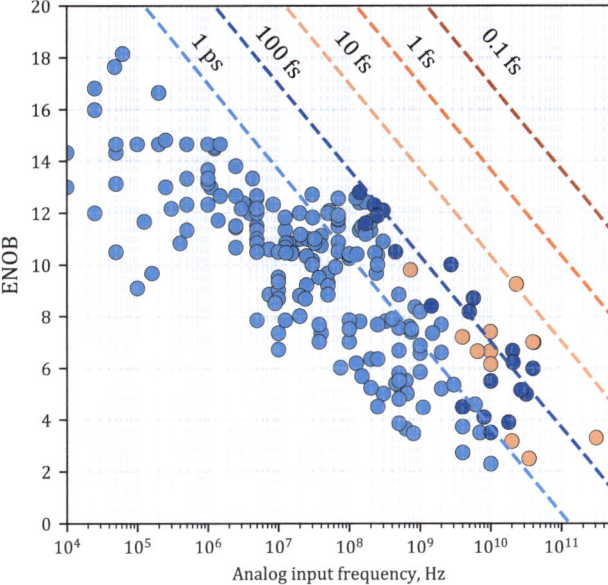

Fig. 5.2 Survey of ADC performance presented as ENOB vs. analog input freqyency (Walden plot). The light blue circles correspond to the original data from the Walden survey [11]. The dark blue circles represent updated data for electronic ADCs, based on the 2024 version of the survey by Murmann [12]. The orange circles denote photonically assisted ADCs from the recent literature, including results from Khilo et al. [13], Ghelfi et al. [14], Xu et al. **(year?)**), Tu et al. [16] and Fang et al. [6]. Dashed lines indicate the ENOB limits imposed by timing jitter levels ranging from 1 ps (blue) to 0.1 fs (red)

The pump light is introduced into the cavity via an integrated wavelength-division multiplexing (WDM) coupler [21], which combines the pump and signal into the gain section. This device is also referred to as a pump-signal combiner. In our implementation, 1610 nm pump light is injected into the thulium-doped amplifier for amplification of signals around a 1890 nm center wavelength. The WDM has a 130 nm bandwidth (at the 1 dB level), while providing over 90% transmission at the pump wavelength. In reverse, the WDM guides the amplified signal to the ACG that closes the cavity for the signal [20].

The signal is amplified in a thin (~ 1 µm) rare-earth-doped alumina gain film, which is deposited on a thin silica top cladding above the silicon nitride (SiN) waveguides. The waveguides have been specifically engineered to support weakly guided modes with large mode area (> 10 µm^2) and a high overlap between the pump and signal within the gain film. More than 86 % of the pump mode is confined within the gain film exhibiting over 99 % mode overlap with the signal. The structure is designed for low-loss, single-mode operation at the pump and signal wavelengths, with higher-order modes being suppressed [22]. Unlike conventional SiN waveguides, which typically have limited mode area, the larger mode area of the gain film provides several critical advantages, including reduced nonlinear effects,

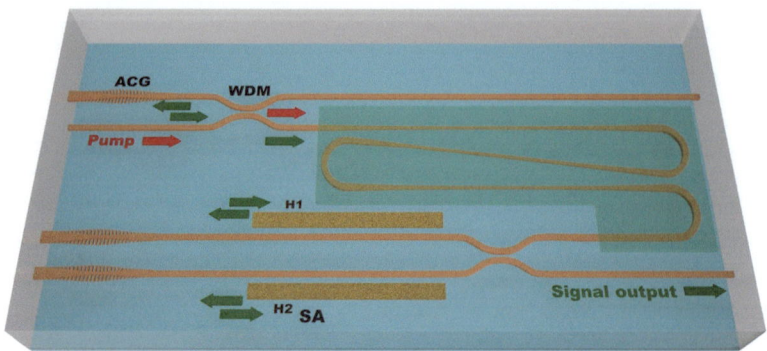

Fig. 5.3 Architecture of the integrated rare-earth MLL. The device is fabricated using a silicon nitride-on-insulator waveguide platform combined with a rare-earth-doped alumina film for optical gain. A wavelength-division multiplexer (WDM) couples the pump light into the gain section. The laser cavity includes a large-mode-area (LMA) waveguide design for high-energy pulse operation, implemented within the region highlighted by the green box. LMA sections are connected through adiabatic tapers and tight bends. Mode locking is achieved using a nonlinear Michelson interferometer-based saturable absorber (NLI-SA), which splits the pulse into two arms via a directional coupler and reflects them back through integrated Bragg mirrors. Thermo-optic phase tuning is implemented via high-power (H1) and low-power (H2) heater arms. Dispersion compensation is achieved via an apodized chirped Bragg grating (ACG)

increased saturation power, larger energy storage and a reduction of thermo-refractive noise. These features are essential for generating low-jitter pulses. Experimental results related to the separately tested amplifier components will be presented in Sect. 5.4.1.

After the gain section, the amplified signal at 1890 nm is transferred back into the silicon nitride core, that is adiabatically tapered up to pull the light back into it. The light then enters a nonlinear Michelson interferometer-based saturable absorber (NLI-SA) with an unequal power splitting ratio between the two interferometer arms. This structure operates as an intensity-dependent reflector and serves as a passive mode-locking element based on the Kerr nonlinearity. For a specific bias phase between the two arms of the interferometer, that can be selected via the heaters, at low incident pulse peak power, the NLI-SA reflects the signal with a significant extinction. At higher input peak powers, the unequal nonlinear phase accumulation in the two interferometer arms leads to increased constructive interference in the reflection path and thus to a higher reflectivity. This mimics the behavior of a saturable absorber and supports pulse formation by preferentially reinforcing the high-intensity portions of the signal. Experimental results validating the nonlinear response and saturable absorption behavior of the NLI-SA will be presented in Sect. 5.4.2.

End mirrors are integral components of the MLL architecture. They define the cavity boundaries and provide optical feedback. One end mirror is positioned at the signal port of the WDM coupler, the others are placed at the extremities of the NLI-arms. They can be implemented as either Sagnac-loops or integrated Bragg gratings. Sagnac loops use waveguide loops and directional couplers, requiring careful design for 3-dB splitting ratios, insertion loss minimization, and broadband operation. In contrast, ACGs offer a compact footprint (typically a few hundred micrometers), broadband reflectivity, and the ability to engineer dispersion within the cavity, which is crucial for ultrashort pulse generation. However, realizing such gratings in a silicon photonics platform poses fabrication challenges, particularly to achieve precise chirp profiles that are compatible with deep ultraviolet (DUV) lithography.

The output signal is directed to the second (reject) port of the NLI. The amount of light that is transmitted out of the cavity is determined by the interference between the signal reflected in the two interferometer arms and the unequal power splitting ratio.

The details of the layer thicknesses and of the device geometries are given in the next section. Simulations and experiments were performed for transverse electric (TE) polarized light for the Q-switched laser and for transverse magnetic (TM) polarized light for the power amplifier described in the following.

5.4 Experimental Results: Key Components and Q-Switched Laser Operation

This section presents experimental results for the main building blocks used in the integrated MLL, including rare-earth-doped amplifiers, a saturable absorber for passive mode-locking and broadband Bragg gratings in silicon nitride technology for dispersion compensation. To date, Q-switched laser operation has been achieved with all these components integrated in a single device fabricated in a silicon photonics platform, with the exception of the ACGs that remain to be integrated with the rest of the components in order to provide the necessary dispersion compensation for stable mode locking.

5.4.1 Rare-Earth-Doped Aluminum Oxide Optical Amplifier for High Power Operation

At the heart of the mode-locked laser lies a rare-earth-doped amplifier section that generates and maintains the optical signal. High pulse energies and ultra-short pulses are desired to achieve low timing jitter, as highlighted in Eq. 5.2. This requires a large mode volume in the gain medium, as the energy stored in the gain medium, E_{stored}, increases with the total number of gain ions N_{tot} that interact with the optical mode: $E_{stored} \propto h\nu N_{tot} = h\nu n_1 L A_{eff}$, in which n_1 is the gain ion density, L is the length of the gain section, A_{eff} is the effective mode area, and $h\nu$ is the energy stored in a single ion which is excited by a pump photon (in this case $h\nu \approx 0.77$ eV). To increase the mode volume, it is more beneficial to increase the mode area rather than the length of the gain medium, because, apart from the benefits of reduced optical nonlinearity, this also increases the saturation energy and saturation power [23], which are given by

$$E_{\text{sat}} = \frac{h\nu_s A_{\text{eff}}}{\Gamma(\sigma_{\text{abs}} + \sigma_{\text{em}})} \quad \text{and} \quad P_{\text{sat}} = \frac{h\nu_s A_{\text{eff}}}{\Gamma(\sigma_{\text{abs}} + \sigma_{\text{em}})\tau}, \tag{5.4}$$

respectively. In these expressions, $h\nu_s$ is the signal photon energy, $\sigma_{\text{abs}} \approx 2 \times 10^{-25}$ m^2 and $\sigma_{\text{em}} \approx 4 \times 10^{-25}$ m^2 are the absorption and emission cross-sections of the gain medium at the pump and signal wavelengths, with numerical values given for the utilized thulium ions, $\tau \approx 1$ ms is the effective lifetime of their excited state, and Γ is the mode confinement factor with the gain material, which can be close to 90 %. A high saturation energy means that the gain medium can support very high-energy pulses before the gain saturates, thus allowing for extraction of high-energy pulses from the amplifying medium.

For this purpose, we implement a hybrid structure in the gain section of the cavity with a rare-earth-doped aluminum oxide (Al_2O_3) gain film deposited on top of the passive SiN waveguide structure. In principle, many rare-earth ions are suitable as dopants in the Al_2O_3 host material, but thulium was chosen in this demonstration due to its broad gain spectrum spanning from 1.8 to 2.0 µm in wavelength, which is ideal for the generation of ultrashort pulses. The design enables a large mode area (LMA) inside the gain layer similar to a weakly guided slab mode, between 20 and 100 µm^2 depending on the exact design. Under these conditions, the mode has a 20 times larger area than a waveguide mode that is highly confined in the SiN waveguide core, which drastically increases the saturation energy. When fully inverted, a rough estimate of the stored energy in the cavity yields around 50 µJ and the saturation energy of the gain medium is in the order of a few µJ, considering a gain section length of 6 cm and a signal around a 1850 nm wavelength, all within a device footprint of less than 10 mm^2. In steady-state mode-locked operation, the pulse would extract every round trip an amount of energy from the gain medium which has to be replenished by the continuous wave pump (~100–500 mW), from which an intra-cavity pulse energy in the order of 1 nJ can be estimated at a ~GHz repetition rate and ~10 % output coupling. Still, the saturation energy gives an estimate of the power-handling capabilities of the LMA amplifier. In fact, as we shall see in Sect. 5.4.4, the intra-cavity pulse energy for such a laser was close to 1 µJ in Q-switched operation, albeit at a 1 MHz repetition rate, demonstrating the high energy storage capacity of the gain section.

To separately test the gain section of the laser device, we fabricated a single-pass amplifier with an identical LMA design as in the laser and thoroughly investigated its amplifying properties [24, 25]. A schematic of the amplifier waveguide structure and its cross-section are shown in Fig. 5.4a and b. The layer stack is fabricated by creating a local opening in the top SiO_2 cladding of the passive waveguide structure, such that only about 300 nm of the SiO_2 remains on top of the SiN waveguide core, that is itself etched into an 800 nm thick stoichiometric SiN layer. A roughly 1 µm-thick thulium-doped Al_2O_3 film is then deposited via reactive RF co-sputtering on top of the chip, such that the local opening is filled with the active gain film. The measured doping concentration is around 4×10^{20} cm^{-3}, which is more than an order of magnitude higher than in typical fiber-based laser and amplifier systems, to achieve high gain within a very short amplifier waveguide. To create a large mode area in the gain film, the SiN waveguide width is reduced to 280 nm in the straight gain sections (located inside the green box in Fig. 5.4), which pushes the optical mode into the gain layer with a confinement factor of over 85 %. Before entering the waveguide bends, the SiN width is slowly tapered back up to 900 nm to create an adiabatic transition pulling the mode back into the SiN core, with this high confinement allowing for a compact bend radius below 85 μm with over 99 % transmission.

A pulsed supercontinuum signal was used for high-power amplification measurements with average signal input power levels up to 36 mW on chip (fiber-to-chip coupling losses of ∼2.4 dB are normalized out). The signal was spectrally filtered from 1.8 to 2.0 μm to match the gain bandwidth of the thulium gain ions and the pulse width at the chip input was approximately 10 ps. A high-power continuous wave pump laser at a 1.61 μm wavelength was used to optically pump the gain medium in a co-propagating pump configuration (signal and pump are coupled to the chip from the same side). The input and amplified output spectra are shown in Fig. 5.5a, showing a strong gain > 10 dB over a bandwidth exceeding 200 nm, with maximum gain achieved around 1850 nm. The amplified signal output power is shown in Fig. 5.5b as a function of pump power and various signal input power levels. On-chip power levels up to 1 W were achieved with roughly 1.5 W of pump power in a 6 cm long amplifier, demonstrating watt-class power-handling capability and a high pump-to-signal power conversion efficiency of ∼66 % in this high-power amplifier operation regime.

Measurements of the small-signal gain have also been performed for a 10 cm long amplifier waveguide with a continuous wave signal at a 1818 nm wavelength and input power levels between 2 μW and 15 mW. On chip small signal net gain was measured to be >28 dB and a high output saturation power was also measured. A high output saturation power means that even high-power signals are still able to achieve high net gain. The output saturation

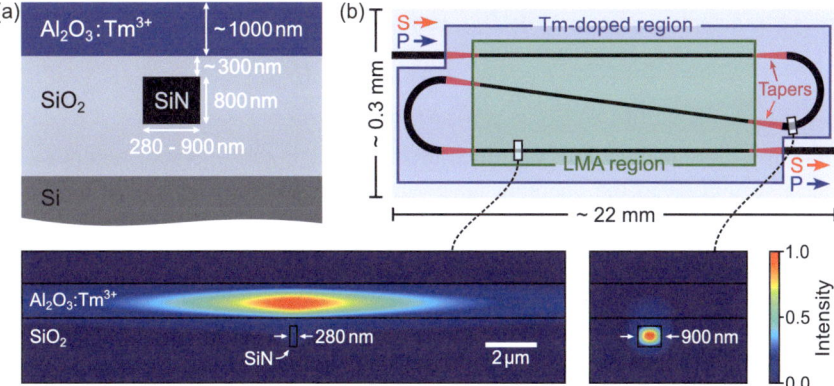

Fig. 5.4 a Cross-section of the gain section of the on-chip amplifier. b Schematic of the amplifier waveguide structure with mode profiles shown for the low-confinement LMA region and for high-confinement bends

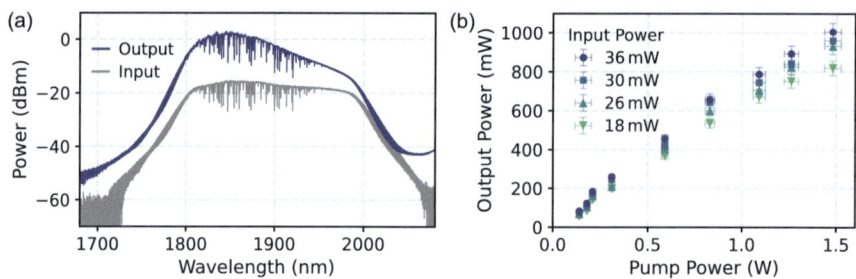

Fig. 5.5 a Input and amplified output spectra of a spectrally filtered supercontinuum pulse. **b** On-chip average amplified output power of the pulsed signal as a function of on-chip pump power for various signal input power levels

power is not identical to the intrinsic gain medium saturation power described by Eq. 5.4, as the output saturation power is also dependent on specific amplifier conditions such as amplifier length and pump power, but in both cases a larger mode area helps to increase the saturation power. The saturation power $P_{\text{sat,out}}$ increases roughly linearly with pump power and reaches beyond 100 mW for the highest pump power used here, which is exceptionally high for integrated optical amplifiers, for which typical saturation power levels in the order of 10 mW have been reported to date.

Moreover, the excess noise added to the signal by the amplifier, which is quantified by the noise figure (NF), was determined from the amplified signal spectra using the optical method [26]. The main source of noise in rare-earth-doped amplifiers is signal-spontaneous emission beat noise, which is the result of the amplified spontaneous emission (ASE) interacting with the amplified signal and degrades the SNR. ASE stems from the spontaneous emission of photons due to the limited lifetime of the rare-earth gain ions, which are then amplified alongside the signal. We determined the NF from the ASE power P_{ASE} within the noise equivalent bandwidth B_0 of the optical spectrum analyzer, the signal photon energy $h\nu_s$, and the gain coefficient on a linear scale G_{lin}, using the relation

$$\text{NF} = \frac{P_{\text{ASE}}}{(G_{\text{lin}} h \nu_s B_0)} + \frac{1}{G_{\text{lin}}} . \qquad (5.5)$$

On-chip NFs as low as 3.7 dB were measured in the high-gain operation regime (>25 dB), which is very close to the quantum noise limit of 3 dB. This highlights the low-noise amplification processes achievable with rare-earth-based gain media and is in stark contrast to prominent semiconductor-based gain media in integrated devices, which are inherently much noisier—semiconductor optical amplifiers usually exhibit NFs of more than 6 dB. Several factors contribute to the excellent noise performance of rare-earth-based gain media. These include the long excited-state lifetime in the order of a few milliseconds, which acts as a low-pass filter for high-frequency external technical noise, such as mechanical vibrations, thermal instabilities, and pump intensity fluctuations. The LMA design further reduces thermo-refractive and thermo-elastic noise, as these are reduced when the total mode volume is increased. This makes the rare-earth-doped LMA gain section suitable for very low-noise pulse generation and amplification.

5.4.2 Nonlinear Michelson Interferometer-Based Saturable Absorber

The operation of the nonlinear Michelson interferometer-based saturable absorber (NLI-SA) is governed by several key parameters. A 2×2 directional coupler splitter (DCS) with an unequal power splitting ratio divides the input signal into two arms with power fractions r and $1-r$. After reflection from integrated mirrors at the ends of each arm, the signals recombine at the same DCS. Some of the recombined light reflects back into the laser cavity while the rest exits through the output port. The reflectivity R of the device is defined as the ratio of reflected energy E_r to input energy E_{in}, and varies depending on the differential nonlinear phase shift $\Delta\phi$ accumulated between the two arms. The reflectivity is maximized near a differential phase shift of π, and the slope of this response determines the saturable absorber coefficient, a key figure of merit for the saturable absorber and its mode-locking driving force. The modulation depth of the artificial absorber, i.e., the variation of the reflectivity over a phase shift of 2π, gives the maximum attainable absorption change. The NLI-SA operates on the upward slope of the first reflectivity peak, where the device exhibits intensity-dependent reflection.

In our implementation, the directional coupler has an 80:20 splitting ratio at a wavelength of 1.9 μm, with more than 50 % reflection modulation depth. The reflectivity curve of the realized device is offset relative to simulations due to imperfections in the fabrication. However, this can be compensated by rebiasing the device with the embedded thermal phase shifters.

There are two main approaches to characterizing the reflection response of an NLI-SA. The first is a direct power-sweep method, where a train of optical pulses is launched into one arm of the NLI, and the average input power is gradually increased. The reflected power is then measured as a function of the launched power. A challenge with this approach arises at higher powers due to significant nonlinear spectral broadening, particularly in waveguides exhibiting anomalous dispersion, which can lead to pulse splitting. As a result, the extracted response is highly dependent on the specific characteristics of the input pulse. A peak pulse power dependent NLI reflection curve has been extracted with a 200 fs pulse width [20].

An alternative method involves keeping the pulse power fixed, typically at low levels to avoid nonlinear effects, while applying an externally tunable phase shift to one of the interferometer arms, for example using an integrated thermal heater. This phase tuning modulates the reflected power without introducing any nonlinear distortion in the input pulse. This approach allows the extraction of a more general NLI response that is independent of the specific pulse shape or peak power and avoids strong self-phase modulation effects present in the first method. Using this technique, multiple π-phase shifts can be mapped out and the maximum modulation depth extracted.

Once the reflectivity curve is known, the device can be thermally tuned to operate in the desired regime. To generate high-energy Q-switched pulses, we bias the device around the highest upward slope. Conversely, operating closer to the base of the slope leads to lower-energy pulses.

Overall, the NLI-SA offers a tunable and fast saturable absorber, with negligible thermal crosstalk between the arms due to their spatial separation.

5.4.3 Broadband Dispersion-Compensating Silicon Nitride Bragg Gratings

For mode-locked operation with femtosecond pulses, careful dispersion compensation must be achieved taking all dispersive elements in the cavity into account. For complete flexibility in dispersion compensation, we designed ACGs operating in the TE polarization with a reflection band centered on 1890 nm. The SiN waveguide is 800 nm thick and has 3.3 μm-thick top and 4 μm-thick bottom oxide claddings. We used deep ultraviolet (DUV) lithography to pattern the Bragg gratings during fabrication, which required us to use fixed corrugation periods. We implemented a linear chirp in the Bragg grating by tapering the waveguide from a narrower to a wider width along the propagation direction, resulting in a linear increase in the effective index. In order to suppress reflection side-lobes and minimize spectral ripples in the dispersion response, we applied a squared-cosine apodization when varying the waveguide width and corrugation depth. The primary design objective was to realize a broadband, flat group delay dispersion (GDD) profile with high reflectivity.

We characterized the fabricated devices by performing transmission and interferometric dispersion measurements. Supercontinuum light in TE polarization was coupled into waveguides with embedded ACGs, and the transmitted and reflected power spectra were recorded with an optical spectrum analyzer. Figure 5.6a shows the measured transmission and reflection spectra obtained from ACGs with 1000 (short) and 2000 (long) Bragg periods, with a grating period of 580 nm and a 50 % duty-cycle. The 3 dB bandwidth of the long grating spans 195 nm. Grating dispersion was measured using a home-built white-light Michelson interferometer. Group delay (GD) and GDD were extracted from a curve fitted to the measured spectral phase response. The GDD level can be accurately controlled through the grating length, with values of $-68,000$ fs^2 and $-140,000$ fs^2 at the center wavelength for the short and long ACG, respectively.

These results confirm that ACGs are capable of supporting broadband dispersion compensation and enabling sub-100 fs pulse formation. Similar gratings centred at 1550 nm were used in separate experiments to demonstrate their suitability for high-performance ultrashort pulse shaping with pulse compression down to 63 fs [27]. Flat dispersion over a wide bandwidth is also crucial for minimizing the time duration of the pulses and therefore timing jitter in MLLs. Our measurements using two different ACG lengths demonstrate both broadband and engineered dispersion control. It is worth noting that the Q-switched integrated laser

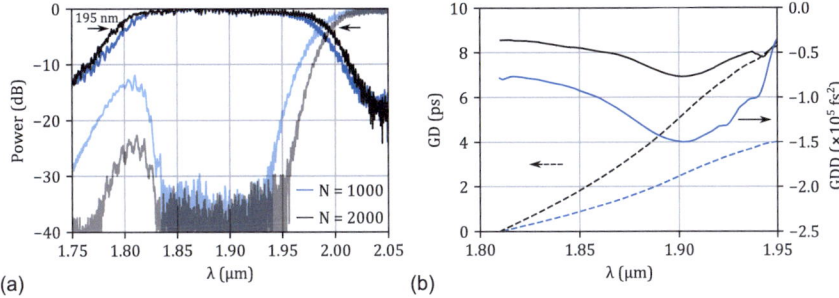

Fig. 5.6 a Measured reflection (full opacity curves) and transmission (reduced opacity) spectra of silicon nitride ACGs centered at 1890 nm, showing high-reflectivity stopbands spanning over 150 nm. The short and long gratings, with 1000 and 2000 Bragg periods, respectively, are shown in blue and black. **b** Extracted group delay dispersion (GDD) and group delay (GD) from white-light interferometry measurements. Solid lines represent GDD (right axis), and dashed lines represent GD (left axis). The GDD curves reveal flat dispersion of approximately $-70,000$ fs^2 for the short grating and $-140,000$ fs^2 for the long grating, showing that the dispersion level can be precisely controlled by adjusting the grating length

described in the next section employed loop mirror reflectors without ACGs, since its design was not yet ready at that time. We believe that integrating ACGs with a taylored dispersion profile is key towards achieving stable mode-locking in future integrated MLLs.

5.4.4 Silicon Photonics High-Energy Passively Q-Switched Laser

Realizing fully integrated, low-jitter MLLs is a critical milestone for enabling next-generation photonic ADCs. In this section, we present a first step into that direction by presenting results on a high-energy, passively Q-switched laser built on a silicon photonics platform [20]. This laser integrates the essential functional components discussed before, especially a rare-earth-doped alumina amplifier, a fast nonlinear Michelson interferometer-based saturable absorber (NLI-SA), and integrated waveguide reflectors. While the system has not yet achieved stable mode-locking, observing energy build-up and pulse formation validates the performance of the amplifier and the function of the saturable absorber within a fully integrated cavity. This result is a crucial step towards on-chip ultrafast sources in which dispersion control, nonlinear dynamics, and gain saturation must be engineered simultaneously in a low-loss silicon photonic technology platform.

The device includes a thulium-doped alumina gain medium with a gain spectrum centered at 1.89 µm and implemented as an LMA waveguide, as discussed before. The mode areas of signal and pump, in the gain sections, are 26.7 µm^2 and 26.3 µm^2, respectively, resulting in >99 % overlap between the two. The confinement factors of the signal mode are 75 % in the aluminium oxide gain film, 0.2 % in the SiN layer, and 11.8 % and 13 % in the top and the bottom silica layers, respectively. The total cavity round-trip length is 16.7 cm, out of which 12 cm correspond to the gain section. The NLI-SA employs a 2×2 DCS with an 80:20 power splitting ratio at 1.89 μm, yielding a modulation depth exceeding 50 %. Unlike the previously discussed dispersion-compensating ACGs, this laser uses loop mirrors as end reflectors in both interferometer arms and at the signal port of the wavelength division multiplexer (WDM).

Q-switched pulses are generated by thermally tuning the NLI-SA so that the intracavity signal is on the positive slope of its reflectivity curve. In this regime, effective loss modulation initiates periodic energy buildup and depletion. A pump at 1.61 μm is coupled into the amplifier section via a lensed fiber. As the pump power increases, the laser transitions from continuous-wave to passively Q-switched operation. This transition is governed by the interplay between gain dynamics and the saturable absorber response. With 400 mW of coupled pump power, the laser emits Q-switched pulses centered at 1.89 μm, which is aligned with the optimal reflection and transmission characteristics of the WDM and loop mirror interfaces. These pulses have a low repetition rate below 1 MHz, a pulse width around 250 ns, and a pulse energy exceeding 150 nJ in the output waveguide of the NLI-SA outside of the laser cavity. The on-chip power conversion slope efficiency is 40 %, with a lasing threshold of about 20 mW. These results confirm stable Q-switched operation and good energy scaling with pump power. Figure 5.7 shows the experimental output, including the optical spectrum, pulse train, and a zoomed-in single pulse.

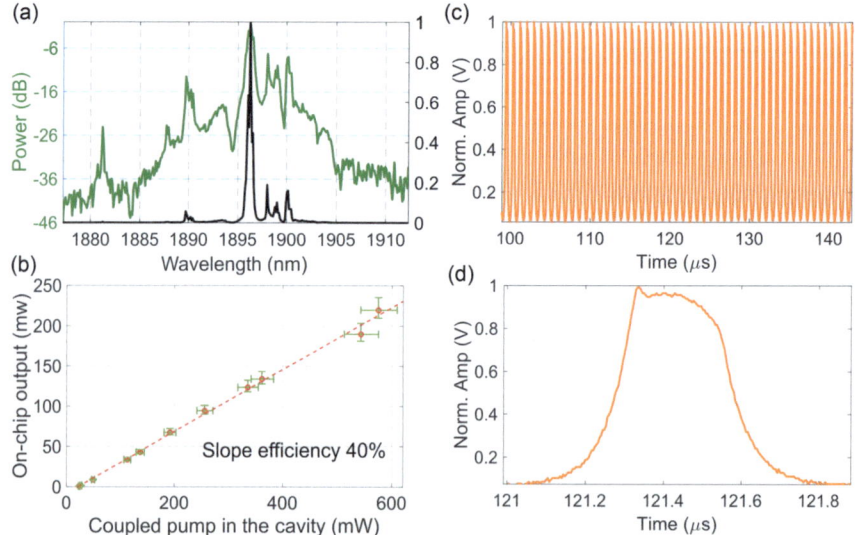

Fig. 5.7 a Measured optical spectrum showing broadband lasing from the integrated cavity. b Average output signal power with respect to the pump power. c Q-switched pulse train recorded with an oscilloscope. d Zoomed-in trace showing the shape of a single pulse

Again, the current laser cavity used loop mirrors without dispersion compensation, which hindered short pulse operation. Future designs can address this issue by incorporating ACGs to engineer dispersion and promote mode locking. Finally, in the Q-switching regime observed here, dynamics must be managed carefully to prevent excessive intracavity power levels that can otherwise damage waveguides. Despite these limitations, demonstrating stable, high-energy Q-switched operation validates the saturable absorber function and confirms the functionality of the fully integrated platform, including the performance of the gain medium.

5.5 Photonic ADCs: Application of Low Jitter Mode-Locked Lasers and Integrated Pulse Shaping

In this section we summarize the architecture of time-interleaved (TI) photonically assisted ADCs, as an example of systems that leverage the low jitter of MLLs to overcome the limitations of purely electronic approaches. We then describe in more detail the implementation of a pulse shaper required for such systems, that is implemented in the same technology platform as the laser and could thus directly be implemented with it.

5.5.1 Photonic ADC Architecture

Photonic ADCs offer three advantages over their purely electronic counterparts: (1) High-precision sampling enabled by ultra-low timing jitter from MLLs, (2) increased effective sampling rates through optical interleaving, and (3) subsequent down-sampling, which enables the use of standard, lower bandwidth electronic ADCs.

The system, whose diagram is represented in Fig. 5.8, builds on an ultra-low-jitter MLL, which generates a train of optical pulses at a repetition rate $f_R = 1/T_R$. These pulses serve as the sampling clock with sub-10-femtosecond timing jitter, and potentially even lower in optimized configurations. To overcome the bandwidth limitations of electronics, the sampling rate is increased by interleaving N pulses with different center frequencies, wherein N is also the number of interleaved channels. The result is a pulse train with an effective repetition rate of $N \cdot f_R$. The electro-optic sampling is performed by modulating this interleaved optical pulse train with the analog RF signal using a broadband Mach-Zehnder intensity modulator. Each optical pulse acquires amplitude information corresponding to the analog signal at the instant of sampling. The resulting modulated pulse train thus encodes the waveform of the input signal as a sequence of intensity-modulated optical pulses. Because the interleaved pulse rate $N \cdot f_R$ may exceed the capabilities of conventional electronic ADCs, the modulated optical signal is demultiplexed according to pulse wavelength into N separate channels using a WDM demultiplexer. This step effectively separates the high-repetition-rate pulses into multiple parallel channels operating at the lower repetition rate f_R. Each

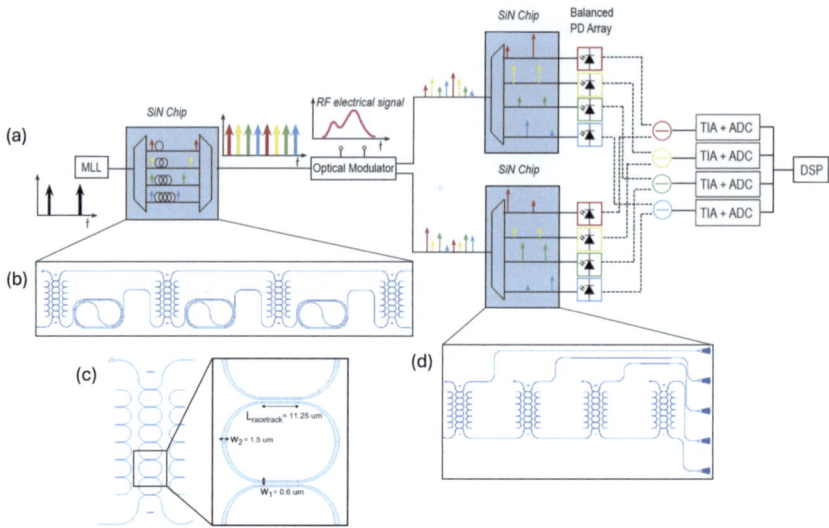

Fig. 5.8 Schematic of the pulse shaper and of its utilization in a photonically assisted TI-ADC. **a** Schematic of a TI-ADC with four channels that allow demultiplying the sampling rate accordingly. **b** Layout of the pulse interleaver and **c** of the coupled resonator optical waveguide (CROW) optical add-drop multiplexer (OADM) used in it. The demultiplexer shown in (**d**) utilizes identical CROW OADMs to facilitate spectral matching. (Figure adapted from [28] under open access license agreement, © 2022 Optica)

lower-bandwidth optical channel is converted into an electrical waveform by a photodetector and is subsequently digitized using an electronic ADC operating at a lower sampling speed, as referenced to the MLL. In addition to reducing the sampling rate requirements of the electrical ADCs digitizing each channel, this also reduces the sensitivity to aperture jitter, as the achievable ENOB can be evaluated with Eq. 5.3 by substituting f_{in} with the frequency of the down-sampled signal actually seen by the electronic ADC, whose maximum value is N times smaller than the Nyquist frequency applying to the overall system.

The photonic TI-ADC architecture offers the advantage that in the ideal case the recorded samples can be directly interleaved to recover the entire signal, without requiring advanced digital signal processing (DSP). However, as a downside, it is also relatively hungry in terms of the spectral comb width required from the MLL, which scales as N^2 when the number of channels is increased while keeping the initial pulse repetition rate of the MLL constant to increase the overall sampling rate of the system. This stems from the fact that interleaving N pulses also requires pulses that are N times smaller and that each require an N times larger spectrum per pulse. The spectral requirements can be reduced by some amount by introducing a multiple-input, multiple-output (MIMO) filter in the DSP [28].

5.5.2 Pulse Interleaver and WDM Demultiplexer for TI-ADCs in the Silicon Nitride Platform

The pulse interleaver is a key component for the integration of a photonically-assisted TI-ADC, following the architecture described in Sect. 5.5.1. It is advantageous if it can be realized in the same SiN platform as the MLL to reduce the number of photonic chips that need to be combined, in particular since the bonding of thin lithium niobate films onto SiN waveguide cores also provides a path towards integration of high-speed modulators into the same chip [29].

Here, we report such a device implemented in the SiN technology of Ligentec SA [30], with waveguides etched into an 800 nm thick stoichiometric Si_3N_4 layer deposited by low pressure chemical vapor deposition (LPCVD) that is compatible with the MLL platform described in the previous sections. For a first proof of principle, the design was targeted for combs centered on 1550 nm [28], but can be straightforwardly adapted to the center wavelength of the integrated MLL.

As a primary advantage for this application, the SiN platform offers a high power handling capability, which is essential to overcome shot noise limited operation in analog signal processing schemes. However, the medium index contrast between SiN and the SiO_2 cladding also limits bending radii and thus routing density, and, in case of ring resonator based devices, the FSR of the filters. Here, we describe a device concept based on coupled resonator optical waveguide (CROW) add-drop multiplexers (OADMs) in which special emphasis was placed onto increasing the filter FSR, and thus the usable spectrum, while minimizing excess optical losses resulting from internal bend-induced resonator losses.

The photonically assisted ADC whose schematic is shown in Fig. 5.8a has four channels and is aimed at being operated with a comb with a 25 GHz repetition rate, resulting in an aggregate sampling rate of 100 GS/s. Figure 5.8b shows the schematic of the pulse interleaver, in which four cascaded CROW OADMs sequentially couple spectral slices, that each have a 360 GHz bandwidth, from an input to an output bus waveguide. Delay loops interposed between one OADM and the next apply 10 ps delays on the input bus and result in the targeted pulse interleaving in the time domain. The OADMs are designed to have an FSR of 1.44 THz, so that four optical pulses can be interleaved in the frequency domain. If the spectrum of the MLL exceeds 1.44 THz, as exemplarily shown in Fig. 5.10b and c, each time-domain pulse will comprise multiple spectral slices separated by one filter FSR, so that the entire power sourced by the MLL is allocated to one of the pulses. This remains compatible with the TI-ADC architecture so long as the cumulative dispersion in the system remains low enough not to separate the different spectral components prior to photodetection and the FSR of the WDM demultiplexer used in the architecture is matched to that of the pulse interleaver. Figure 5.9 shows a micrograph of the device.

Figure 5.8c shows a detailed view of the CROW OADM layout and Fig. 5.10a a comparison between the modeled and measured CROW OADM transfer functions. There can be seen to be a very good agreement between the two. Adjacent OADMs are also designed to have passbands shifted by 360 GHz by design, which is achieved by adding small waveguide increments in the layout of the individual resonators forming the CROWs. Figure 5.10b shows the transfer function of the four CROW filters, that have been individually measured without thermal tuning, which also shows that they have the targeted relative spectral alignment as fabricated. Due to the periodic nature of their transfer function, their absolute spectral position is also not critical so long as the transfer functions of the CROWs utilized in the pulse interleaver and the nominally identical CROWs used for WDM demultiplexing of the signal after modulation (see Fig. 5.8d) spectrally overlay with each other. In other words, the repeatability of the process may be sufficient to operate this device without any thermal tuning, which is very desirable to reduce its power consumption in view of the low thermo-optic coefficient of SiN.

In order to obtain a sufficiently steep transfer function for the targeted system, sixth order CROW filters were designed that incorporate each sixth racetrack resonators coupled to each other. The coupling strengths, chosen to implement a Van Vliet transfer function selected for its flat group delay [31], were set by varying the gap between one resonator and the next. The resonators themselves were implemented as racetracks, with the waveguides kept thinner in the straight sections to increase the coupling strengths (0.6 μm waveguide width, 11.25 μm coupling section length), and adiabatically increased to 1.5 μm in the bends (12.2 μm radius) to reduce the bending losses. This design approach allowed minimizing the circumference of the resonators, by keeping the straight sections short and reducing the bending radii, so as to achieve the targeted 1.44 THz filter FSR, while maintaining low bending losses and the high coupling strengths required to obtain a 360 GHz passband. The insertion losses of the CROW filters, 3.5 dB at the center of their passband, are also in close agreement with simulation results assuming waveguide losses to be 0.3 dB/cm.

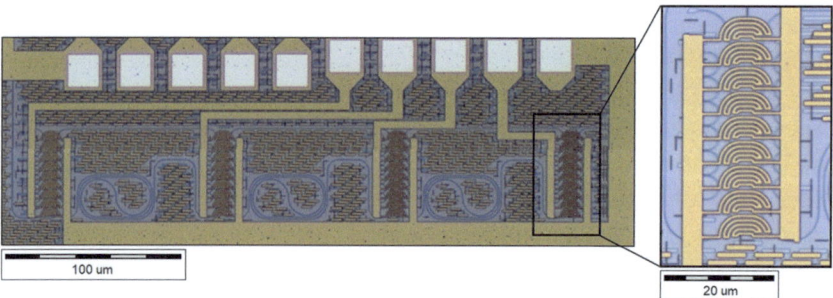

Fig. 5.9 Micrograph of the pulse interleaver with four CROW OADMs and three interspaced delay lines. The inset shows a detailed view of the metal heaters fabricated on top of the rings for thermal tuning. These share common electrodes and allow for a global spectral repositioning of the OADM. The individual rings inside the CROW are spectrally aligned relative to each other by design and do not require individual tuning. (Figure adapted from [28] under open access license agreement, © 2022 Optica)

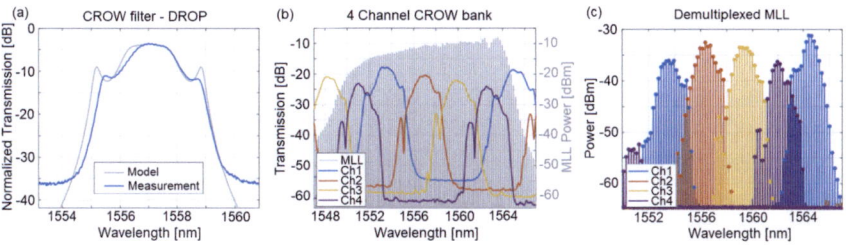

Fig. 5.10 Transfer functions of the CROW OADMs. **a** Overlay of the simulated and measured transfer functions of a single OADM. **b** Measured transfer functions of the four CROW OADMs, without any thermal tuning, and overlay with the spectrum of a semiconductor MLL [32, 33]. The transfer functions can be seen to be staggered in wavelength, as designed. The partial frequency overlap was designed in and results from a trade-off between pulse width (sampling accuracy) and frequency domain cross-talk [28]. **c** Semiconductor MLL spectrum exemplarilly filtered by the CROW filters. It can be seen that the pulses color-coded in blue and in purple each comprise two frequency slices separated by one filter FSR. This is not a problem as long as the cumulative dispersion in the system stays low enough for these spectral components not to separate. (Figure adapted from [28] under open access license agreement, © 2022 Optica)

5.6 Cascaded Optical Filters for Spectral Shaping of Integrated Mode-Locked Lasers

With the explosive growth of data transport and processing worldwide, increasing strain is placed on communication networks to meet demand. In this context, optical frequency comb sources (OFCs) are investigated as a means to increase network capacity [33, 34]. The large number of equally spaced comb lines that an OFC generates are ideal for communication systems, since each line can serve as an optical carrier on a perfectly aligned grid to allow massively parallel WDM transmission. OFCs are also instrumental for distance ranging [35], signal synthesis [36], neuromorphic computing [8, 37, 38], and a class of photonic ADCs that perform the optical slicing and DSP in the frequency domain [6, 7].

For these applications, the FSR of the comb needs to be in the order of tens of GHz, since this determines the bandwidth of the tributary signals that can be applied to the comb lines. This is readily achieved with semiconductor MLLs [32] or with on-chip parametric comb generation [39]. However, it is currently out of reach of integrated rare-earth MLLs, due to the extended cavity size required due to the comparatively low gain coefficient of the rare-earth gain media.

The pulse repetition rate can be increased in the time domain with a passive pulse interleaver in a lossless manner [40]. However, this results in the interleaved pulses having different phases and does not increase the FSR in the frequency domain. An elegant extension of this method has been shown in [41], in which the phase of the interleaved pulses has been dynamically corrected with a phase shifter, resulting in an actual increase of the FSR. This does, however, require a high-speed modulator to apply phase shifts that need to be adjusted as the temperature of the pulse interleaver drifts with an update rate that needs to be synchronized with the MLL's pulse repetition rate.

We present a simpler to operate and more straightforward approach consisting in designing a higher-order filter with multiple optical ring resonators (ORRs) implemented in the same SiN platform as the MLL described in the previous sections. Such optical filters have already been integrated in silicon photonics platforms to select groups of comb lines [42]. Here, an additional challenge arises from the OFC starting with a relatively small FSR, assumed in this section to be 2.5 GHz, which makes it difficult to isolate individual comb lines in an integrated platform with finite waveguide losses that limit achievable quality (Q-)factors and thus resonator linewidths. While such filtering has been implemented with fiber-based ORRs starting from even smaller FSRs [43], the 0.2 dB/cm waveguide losses typical for high-confinement SiN plaforms limit loaded Q-factors to $\sim$1,000,000 and linewidths to a $\sim$200 MHz FWHM, assuming a comb centered at 1550 nm. This in turn limits the rejection of adjacent comb lines to less than 30 dB, which is further reduced to $\sim$20 dB extinction if the initial FSR is reduced to 1.25 MHz, which corresponds to the laser cavity design presented in Sect. 5.4.

Operating ORRs as OADMs, that pick every Nth line from the input bus waveguide and couple them to a drop waveguide, exacerbates this problem, as there exists an inherent trade-off between the insertion losses (ILs) of OADMs and their linewidth resulting from the loaded Q-factor: In order to have low ILs, the extrinsic resonator losses corresponding to coupling to the bus waveguides need to be significantly larger than the internal losses arising from the waveguide losses inside the ring. This in turn spoils the Q-factor of the ORR and the ability to select lines from a closely spaced OFC.

We circumvent this problem by cascading two ORRs. One is operated as a higher-Q notch filter in critical-coupling configuration, that extinguishes every second comb line and whose loaded Q-factor is half of the intrinsic Q-factor provided by the platform. This doubles the FSR and reduces the Q-factor required from the other ORR, that is operated as an OADM picking up every Nth comb line of the original comb and whose Q-factor can be sufficiently spoiled to enable low ILs.

This approach can be configured to upconvert the FSR of the MLL to any even multiple, e.g., FSRs of 5, 10, 20 or 25 GHz can be obtained this way from the original 2.5 GHz. While this approach seems deceptively simple on paper, a practical implementation requires a precise adjustment of the coupling strengths to achieve the targeted extinctions as well as a precise control of the ORR's FSR and waveguide dispersion, as these otherwise limit the spectral width over which the OFC can be effectively filtered without incurring large excess losses due to spectral misalignment between comb and filter.

An overview of relevant resonator theory is given in Sect. 5.6.1, followed by the design details in Sect. 5.6.2 and first experimental results in Sect. 5.6.3.

5.6.1 Theoretical Background for Ring Based Filters

ORRs are fundamental building blocks for infinite impulse response filters in integrated photonics [44]. Generic ring resonator filters in all-pass and OADM configuration are depicted in Fig. 5.11.

The transfer functions of the OADM, wherein the all-pass configuration can be seen as a special case with $t_2 = 1$ ($\kappa_2 = 0$), can be derived as

$$E_{thr} = \frac{t_1 - e^{(i\beta - \frac{\alpha}{2})L} t_2}{1 - e^{(i\beta - \frac{\alpha}{2})L} t_1 t_2} \quad (5.6)$$

$$E_{drop} = \frac{-\kappa_1 \kappa_2 e^{i\phi}}{1 - e^{(i\beta - \frac{\alpha}{2})L} t_1 t_2} \quad (5.7)$$

where E_{thr}, E_{drop} are the field amplitudes at the through and drop ports assuming an input amplitude of 1, $\kappa_{1,2}$ and $t_{1,2}$ are the amplitude coupling and transmission coefficients at the two junctions, L is the roundtrip length of the ORR, β, α are the wave-number and linear loss coefficient of the waveguide, and ϕ is the phase accrued by the light as it transmits from

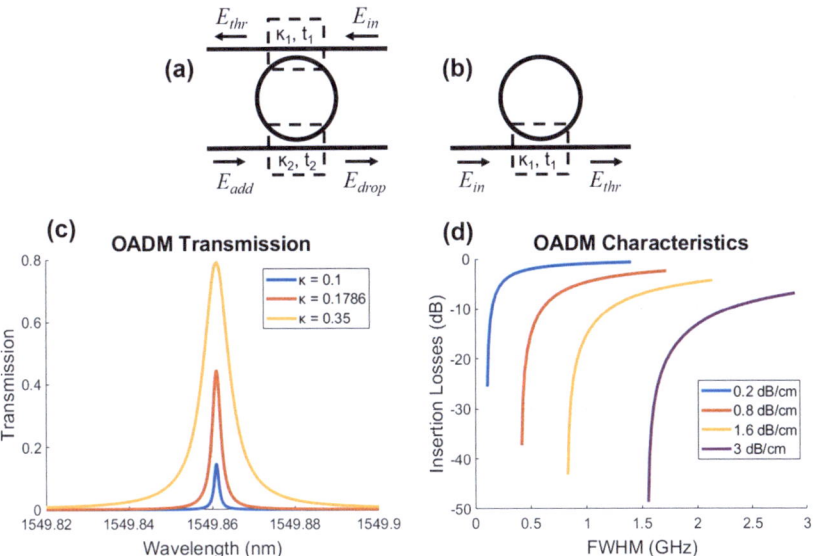

Fig. 5.11 Schematic of **a** an OADM with two bus waveguides and **b** an all-pass ORR used as a periodic notch filter. **c** Transmission spectra at the drop port of an OADM with a 20 GHz FSR for different coupling coefficients, at fixed waveguide losses assumed to be 0.2 dB/cm. A trade-off between linewidth and ILs is apparent. **d** ILs versus linewidth assuming different waveguide losses. The coupling coefficients κ_1 and κ_2 are assumed to be equal to each other and are intrinsically varied inside the curves

the first to the second junction inside the ORR (the corresponding losses incurred in less than a single roundtrip are assumed to be negligible here).

From these, we can extract the essential metrics of the filters, that are the FWHM of the resonance, the FSR of the filter, the extinction of the notch filter, and the ILs of the OADM, according to

$$\text{FWHM [Hz]} = \frac{(1 - t_1 t_2 e^{-\frac{\alpha}{2}L})c_0}{\pi n_g L \sqrt{t_1 t_2 e^{-\frac{\alpha}{2}L}}} \tag{5.8}$$

$$\text{FSR [Hz]} = \frac{c_0}{n_g L} \tag{5.9}$$

$$\text{IL [dB]} = -10 log_{10}\left(\frac{\kappa_1^2 \kappa_2^2}{(1 - e^{-\frac{\alpha}{2}L}t_1 t_2)^2}\right) \tag{5.10}$$

where n_g is the group index of the waveguide and c_0 the speed of light in vacuum. As explained above, at fixed waveguide losses α determined by the technology platform, the trade-off between the FWHM and the ILs can be dialed in by choosing $\kappa_{1,2}$ accordingly. The FSR is required to stay constant across the spectral width of the targeted filtered comb, with cumulative tolerances determined by the filter FWHM (so that filtered or dropped comb lines remain within the linewidth of the resonances at the extremities of the spectrum).

5.6.2 Filter Design

The proposed filter relies on cascading multiple rings [45]. Its schematic is shown in Fig. 5.12. It consists of an OADM cascaded with a notch filter applied to its drop port. The OADM selects the targeted comb lines and couples them to its drop port, while the cascaded notch filter further suppresses lines that are immediately adjacent to the selected ones. Since the notch filter does not need to couple light to a drop waveguide, it can be operated in critical coupling condition with a higher loaded Q, thus increasing the extinction with minimal increase of the ILs at the targeted lines. Its FSR is targeted to be 2× that of the MLL.

For a 20 GHz filter and assuming waveguides with 0.2 dB/cm losses and a group index of $n_g = 2.12$, we target coupling coefficients of $\kappa_1 = \kappa_2 = 0.179$ for the OADM and $\kappa_1 = 0.349$ for the notch filter, with ORR circumferences of $L = 7075$ μm and $L = 28302$ μm, respectively. The intrinsic Q-factor of the ORRs, determined by the waveguide losses, is 1.87 Million. The loaded Q-factor of the critically coupled notch filter is half, 933,000, resulting in a FWHM of 207 MHz. The OADM on the other hand has a reduced Q-factor of 622,000 and a FWHM of 311 MHz. This results in 3.45 dB ILs from the OADM and negligible excess ILs of <0.01 dB from the notch filter at the targeted comb lines. Without the notch filter, the OADM would extinguish their nearest neighbors, that are spaced by 2.5 GHz from them, by 24 dB relative to them, while the second nearest rejected comb lines are extinguished by 30 dB. By using the notch filter, an extinction of >30 dB can be obtained throughout.

This approach introduces several implementation challenges. As mentioned above, precise control over the group index and minimization of the dispersion are essential to ensure that a wideband OFC can be filtered. Moreover, care has to be taken in the layout to ensure that the larger FSR of the OADM is an exact multiple of the notch-filter's FSR, so that they co-vary across process variations. This facilitates guarding against them with a device parameter exploration, since it reduces the dimensionality of the parameter space to one.

To reduce the footprint of the device given the large ORR circumferences required to reach the targeted FSRs, the resonators are realized using spiral waveguide geometries. Each ORR consists of two spirals connected by waveguides in the center region of the ORR, that also comprises the coupling sections, as seen in Fig. 5.13.

Given the large number of bends involved, particular attention was paid to minimizing polarization and modal crosstalk, which can occur due to bends since the waveguides are etched with slightly slanted edges that break their vertical symmetry. A square spiral configuration was adopted, in which the majority of the cavity consists of straight waveguide segments connected via compact 90° bends with a radius of 25 μm. While the resulting bend-

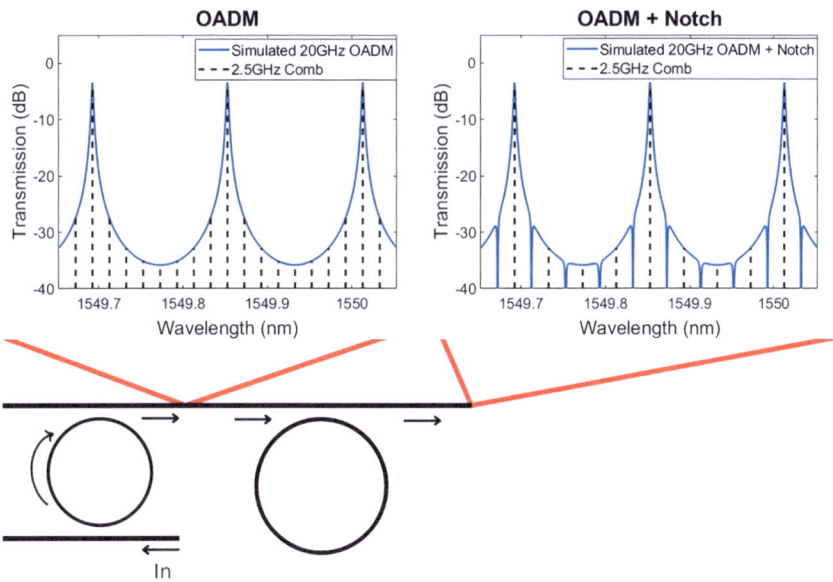

Fig. 5.12 Schematic of the OFC filter. The OFC is applied as an input to the OADM and is filtered as seen on the left plot. The signal is then filtered by the notch filter, with the resulting spectrum shown in the right plot. This illustration is based on a 20 GHz OFC-filter and a 2.5 GHz initial comb spectrum as discussed in the text

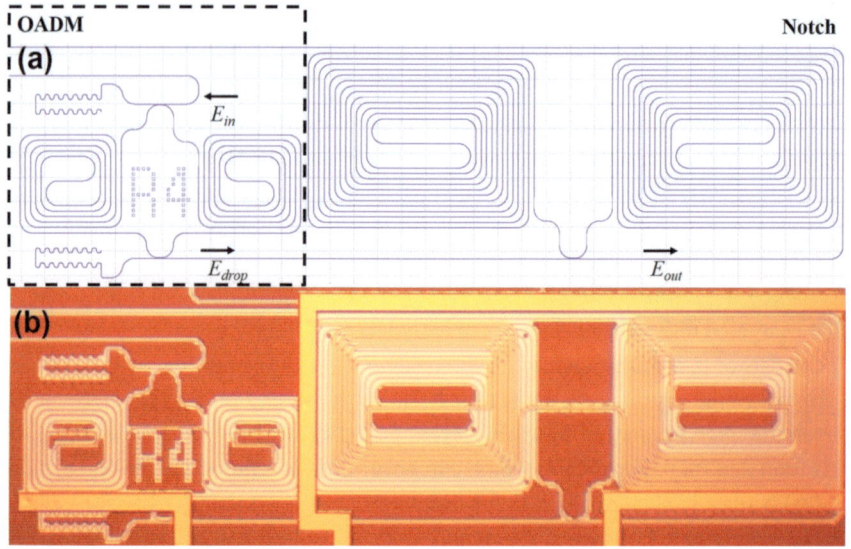

Fig. 5.13 a Layout of a filter converting the FSR of an OFC from 2.5 to 20 GHz. The ports of the device are labeled consistently with Fig. 5.11a and b. **b** Micrograph of the device

ing losses are small for the fundamental TE and TM modes (both below 0.003 dB/cm), the bends introduce significant losses for higher-order modes, contributing to modal filtering.

Unfortunately, these narrow bends also create substantial losses due to modal mismatch between the straight and bent sections that was underestimated during the design phase and amount to 0.019 dB per transition. Given the large number of bends, the cumulative losses in the 20 GHz OADM are about 2 dB per roundtrip, or ∼3 dB/cm. This significantly degrades the performance of the OADM as explained in the experimental section and needs to be addressed by implementing adiabatic bends, such as e.g. Euler bends [46], in the future.

As explained above, the dispersion of the waveguide needs to be minimized to maintain a constant ORR FSR across the comb, see Eq. 5.9. For filtering of a broadband comb centered on 1550 nm, the dispersion was studied in the range of 1540–1560 nm. Over this wavelength range, the Ligentec SiN platform exhibits the smallest dispersion for a waveguide base width of 1.05 µm given the fixed 800 nm waveguide height. At this width, the waveguide is no longer single mode and supports higher-order modes that are very weakly guided. A short section in which the waveguide width was tapered down to 0.8 µm in a bend was introduced as a modal filter in each ring to further suppress the build-up of higher-order modes. The ratio of this single-mode section to the total cavity length was kept consistent across different ring sizes to preserve the targeted ratio between the FSRs.

Fabrication induced shifts in resonance wavelengths are straightforward to correct with thermal phase shifters, as implemented and seen in Fig. 5.13b in the form of metal heater spirals arranged on top of the waveguide spirals. However, the FSR of the ORRs is much

5 Integrated Low Jitter Mode-Locked Lasers and Pulse/Spectrum Shapers

harder to tune and should ideally be correct by design. The group index, n_g, is very sensitive to the geometry of the waveguides, so that even small errors in the fabrication can cause shifts in the FSR dispersion profile that are significant given the large targeted comb spectra and the narrow linewidth of the resonances.

A sensitivity analysis is conducted to quantify this and its results are visualized in Fig. 5.14 after multiplying the spectral response of a filter modeled under consideration of process biases with a train of evenly spaced Dirac peaks emulating a 2.5 GHz OFC. Figure 5.14a shows the attenuation of the targeted comb lines across a 20 nm spectrum while Fig. 5.14b shows the power attenuation of the adjacent comb lines that should be rejected. Deviations which change the cross-sectional aspect ratio of the waveguide, i.e., with opposite offsets applied to the height and to the width of the waveguide core, cause the biggest degradation in performance. Selecting the 3-dB bandwidth as a performance metric, independent variations beyond ± 4 nm in height and ± 10.4 nm in width are not acceptable if a 20 nm comb width is to be maintained. Additionally, these deviations do not only reduce the magnitude of the targeted comb lines but also raise the level of the adjacent ones, reducing the overall extinction ratio. For the range described above, the comb line suppression worsens by ~ 14 dB, rising from ~ 36 dB below the selected comb line transmission to ~ 20 dB below it.

An additional penalty can arise from changes in the directional coupler geometry. In the corresponding analysis, the center-to-center distance between the waveguides is assumed to remain constant and the gap between them to increase or decrease by 10.4 nm, in line with the process compliance limits discussed above for the waveguide width. At the same time, the layer thickness is also varied by ± 4 nm, in a complete corner analysis. This results in small changes of the coupling coefficients, with κ varying in the range of 1–2 %, too small a change to significantly impact the OADM performance (IL changes are ~ 0.1 dB). Provided the fabrication biases also remain inside these bounds in the junctions, the dominant

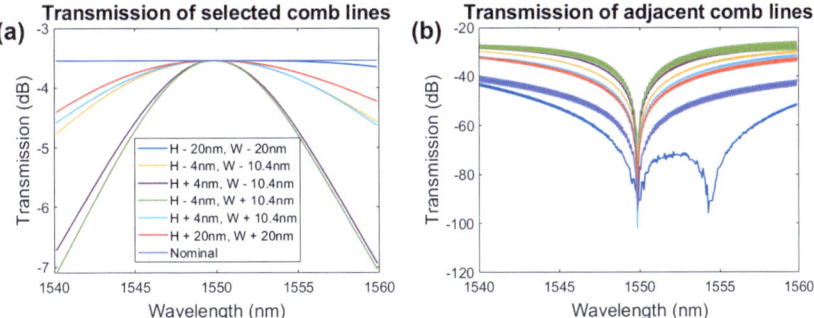

Fig. 5.14 Tolerance to fabrication errors. **a** Attenuation of the selected comb lines versus wavelength for different fabrication biases. **b** Attenuation of the adjacent comb lines on either side of the OADM resonance versus wavelength for different fabrication biases

performance sensitivity would lie in the effective filter bandwidth resulting from the FSR variation. However, different and larger process biases are possible in the junctions.

5.6.3 Experimental Results

We have characterized a device fabricated in the SiN platform of Ligentec SA that was aimed at decimating an OFC to convert its FSR from 2.5 to 20 GHz.

The spectral shaping capabilities of the device was measured using an ultra-stable 2.5 GHz FSR MLL from Menhir Photonics SA and a high-resolution optical spectrum analyzer (OSA) from APEX Technologies with a 5 MHz resolution bandwidth. To thermally align the filter resonances to the grid of the OFC, amplified spontaneous emission generated by an erbium doped fiber amplifier (EDFA) was first combined with the OFC and injected into the device, so that the overlay of the OFC with the resonances could be observed on the OSA.

Post-fabrication metrology revealed that the waveguides were significantly higher (818 nm) and thinner (1.01 µm) than assumed in the design, so that the FSR and coupling strengths were significantly off-target in the fabricated device, beyond the range anticipated in the sensitivity analysis in Sect. 5.6.2. At room temperature, an FSR of 20.038 GHz was measured, in line with simulations once the process bias and the laid out resonator circumference of $L = 7043$ µm were accounted for. This could be corrected by heating up the device to $\sim$200°C on a hot plate, after which the spectrum shown in Fig. 5.15 was measured with the Menhir Photonics laser used as a light source.

After the OADM, the extinction of the lines that are immediately adjacent to the targeted ones, relative to them, is in the order of $\sim$11 dB in a 16 nm range spanning from 192.5 to 194.5 THz (100 target comb lines with a 20 GHz spacing). The lines that are second neighbors to the targeted ones are extinguished by approximately 18 dB. Both numbers are consistent with loaded OADM Q-factors in the range of 130,000–150,000, which are much lower than targeted due to the interface losses between straight and bent waveguide segments that amount to $\sim$3 dB/cm and will be addressed in the future with adiabatic bends. In this configuration, the combination with a functional notch filter appears even more critical to further extinguish the nearest neighbors.

The ILs of the OADM, in the order of 21 dB, are also limited by the below target coupling strengths. Test structures implemented to independently measure them show that the directional couplers have coupling strengths very significantly below target, with measured amplitude coupling coefficients of $\sim$0.105 and $\sim$0.18 versus targeted ones of 0.179 and 0.349, for the OADM and notch filter, respectively. Since this discrepancy exceeds by far that expected from the process bias as evaluated from stand-alone waveguides, it is possible that a different and more significant process bias occurred for closely spaced waveguides in the coupler sections that will require a retargeting of the junctions in the future.

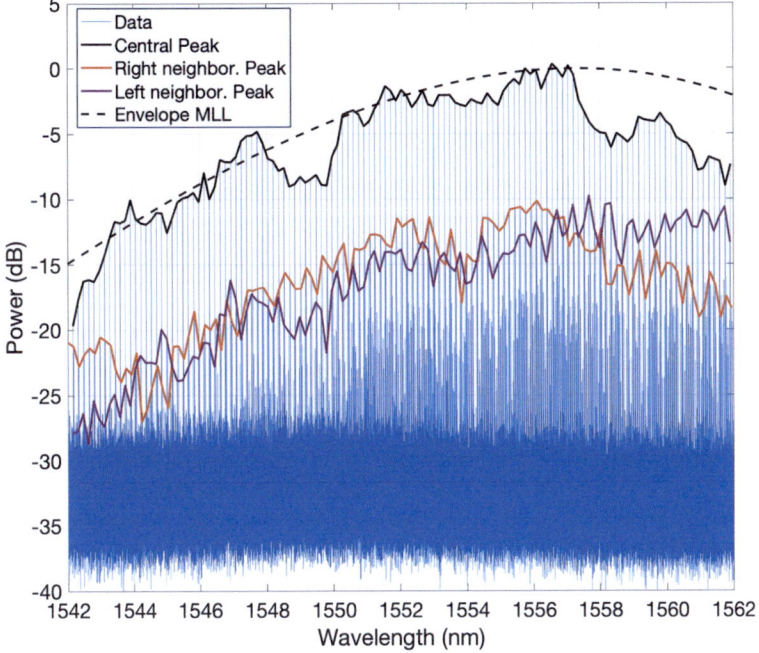

Fig. 5.15 Optical spectrum of an OFC measured over a 20 nm wavelength range after filtering by a 20 GHz comb decimation filter (thin blue curve). The solid black, red, and purple curves show the amplitude of the targeted comb lines, their nearest neighbor towards larger wavelengths, and their nearest neighbor towards smaller wavelengths, respectively. The spectrum of the unfiltered MLL used to perform the measurement is overlaid as a dashed black curve. The unfiltered MLL spectrum and the filtered MLL spectrum are both normalized to have a peak power of 0 dB. ILs are reported separately in the text

At the edges of the recorded optical spectrum, at 1542 and 1562 nm, the nearest neighbor comb lines directly adjacent to the targeted ones and located towards the center of the spectrum, i.e., the higher wavelength ones at 1542 nm and the lower wavelength ones at 1562 nm, are almost as strong after the filter as the targeted ones. This points to the average FSR being below target. We evaluate the resonance frequencies of the OADM to be shifted by a cumulative amount of +1 GHz at 1562 nm (192 THz) and by −1 GHz at 1542 nm (194.5 THz), indicating that the average FSR is 16 MHz too small. This is attributed primarily to the difficulty of achieving precise temperature control at 200°C, however dispersion might also play a role here, since the data indicates that the FSR offsets grow disproportionally towards the edges of the spectrum.

This experimental realization of the filter concept serves as a first demonstration, but also highlights the aspects that require attention in a future fabrication iteration. The 200°C tuning temperate required to adjust the FSR is excessive, but can be very significantly reduced by retargeting the design. Even with the observed room temperature FSR offset, broadband alignment of the filter with the OFC was possible. A future device design will also have to address excessive internal ORR loss by implementing adiabatic bends.

Based on the sensitivity analysis reported in the previous section, we expect that process control of ±4 nm in waveguide height and ±10.4 nm in waveguide width will enable filters with ILs of 3.5 dB, a 3-dB passband of 20 nm, and an extinction above 21 dB for the rejected lines relative to the targeted ones across the entire spectrum (which would be above 30 dB in the absence of process variations).

5.7 Conclusion and Outlook

This work presents key building blocks for low-jitter, mode-locked lasers integrated in a silicon photonics platform. The rare-earth-doped amplifier design allows for a large mode area within the gain film of up to 100 μm^2 and saturation energies of about 5 μJ, all within a footprint of less than 10 mm^2. We also developed and characterized dispersion-compensating, apodized and chirped Bragg gratings with a reflection bandwidth of over 150 nm and a group delay dispersion engineered to be flat across that spectrum. Additionally, we demonstrated a passively Q-switched laser fully integrated in this platform. This laser uses a large-mode-area waveguide amplifier, a nonlinear Michelson interferometer-based saturable absorber, and loop mirrors. It features pulse energies of over 150 nJ in the output waveguide outside of the laser cavity with pulse widths of 250 ns at 1.89 μm. This represents an improvement of $\sim$20 dB in pulse energy and a $\sim$40 % slope efficiency, which is well above previously reported values for integrated Q-switched lasers.

Although stable mode-locked operation is the ultimate goal, several factors limited its realization in this configuration. First, the effective modulation depth of the saturable absorber was low because it was operating near the peak of the NLI reflectivity curve. This resulted in sub-percent gain/loss modulation, favoring long-pulse Q-switched behavior over the formation of ultrashort pulses. Second, the cavity used loop mirrors without dispersion compensation. Such Q-switched dynamics must be managed carefully to prevent excessive intracavity powers that could damage waveguides. Despite these limitations, these results validate the core building blocks of the integrated laser. Future iterations that incorporate dispersion-compensating Bragg gratings are expected to enable the stable generation of femtosecond pulses, which is essential for photonically enabled signal sampling.

In addition to the mode-locked laser itself and its key components, this chapter reports the implementation of pulse interleavers, WDM (de)multiplexers and comb decimation filters implemented in the same platform, that can be directly integrated with the laser and serve the implementation of electrical-optical signal processing systems such as photonic ADCs. Coupled resonator optical waveguide filters with a passband of 360 GHz, a large free spectral range of 1.44 THz and low insertion losses of 3.5 dB are implemented as an essential building block for four channel integrated time-interleaved photonic ADCs.

In addition, a first iteration of comb decimation filters aimed at increasing the free spectral range of a broadband optical frequency comb from 2.5 to 20 GHz are reported, with a design study indicating a 3-dB bandwidth of 20 nm, insertion losses of 3.5 dB and an extinction of unwanted comb lines of over 20 dB to be reliably achievable within typical process variations. The first design iteration will be improved in the future with adiabatic bends.

Acknowledgements The authors would like to acknowledge funding from the German Research Foundation (Deutsche Forschungsgemeinschaft-DFG) for project "Ultra-Wideband Photonically Assisted Analog-to-Digital Converters" (403188360) in the framework for the Priority Programme 2111 "Electronic-Photonic Integrated Systems for Ultrafast Signal Processing", from the German Federal Ministry for Research and Education for project "Scalable Photonic Neuromorphic Circuits", Grant 03ZU1106BA, in the framework of the Cluster4Future NeuroSys, and from the Horizon Europe Programme under Grant Agreement No. 101159229 (femto-iCOMB). The authors would also like to thank Menhir Photonics SA for their support.

References

1. Walden R (2008) Analog-to-digital conversion in the early twenty-first century. John Wiley & Sons Inc
2. Haus HA, Mecozzi A (1993) Noise of mode-locked lasers. IEEE J Quantum Electron 29:983–996
3. Benedick AJ, Fujimoto JG, Kärtner FX (2012) Optical flywheels with attosecond jitter. Nat Photonics 6:97–100
4. Wang Z, Gasse KV, Moskalenko V, Latkowski S, Bente E, Kuyken B, Roelkens G (2016) A III-V-on-si ultra-dense comb laser. Light Sci Appl 6, Article ID e16260
5. Hauck J, Zazzi A, Garreau A, Lelarge F, Moscoso-Mártir A, Merget F, Witzens J (2019) Semiconductor laser mode locking stabilization with optical feedback from a silicon PIC. J Lightwave Technol 37:3483–3494
6. Fang D, Drayss D, Peng H, Lihachev G, Füllner C, Kuzmin A, Marin-Palomo P, Kharel P, Wang RN, Riemensberger J, Zhang M, Witzens J, Scheytt JC, Freude W, Randel S, Kippenberg TJ, Koos C (2025) 320 GHz photonic-electronic analogue-to-digital converter (ADC) exploiting Kerr soliton microcombs. Light Sci Appl 14, Article ID 241
7. Zazzi A, Das AD, Hüssen L, Negra R, Witzens J (2024) Scalable orthogonal delay-division multiplexed OEO artificial neural network trained for TI-ADC equalization. Photonics Res 12:85–105
8. Zazzi A, Müller J, Ghannam I, Battermann M, Vasudevan Rajeswari G, Weizel M, Scheytt JC, Witzens J (2022) Wideband SiN pulse interleaver for optically-enabled analog-to-digital conversion: a device-to-system analysis with cyclic equalization. Opt Express 30:4444–4466
9. Lee TH, Hajimiri A (2000) Oscillator phase noise: a tutorial. IEEE J Solid-State Circuits 35:326–336
10. Kim J, Kärtner FX (2010) Attosecond-precision ultrafast photonics. Laser Photonics Rev 4(3):432–456
11. Walden R (1999) Analog-to-digital converter survey and analysis. IEEE J Sel Areas Commun 17(4):539–550
12. Murmann B (2025) ADC performance survey 1997-2025. https://github.com/bmurmann/ADC-survey
13. Khilo A, Spector SJ, Grein ME, Nejadmalayeri AH, Holzwarth CW, Sander MY, Dahlem MS, Peng MY, Geis MW, DiLello NA, Yoon JU, Motamedi A, Orcutt JS, Wang JP, Sorace-Agaskar

CM, Popović MA, Sun J, Zhou G-R, Byun H, Chen J, Hoyt JL, Smith HI, Ram RJ, Perrott M, Lyszczarz TM, Ippen EP, Kärtner FX (2012) Photonic ADC: overcoming the bottleneck of electronic jitter. Opt Express 20(4):4454–4469
14. Ghelfi P, Laghezza F, Scotti F, Serafino G, Capria A, Pinna S, Onori D, Porzi C, Scaffardi M, Malacarne A, Vercesi V, Lazzeri E, Berizzi F, Bogoni A (2014) A fully photonics-based coherent radar system. Nature 507(7492):341–345
15. Xu S, Zou X, Ma B, Chen J, Yu L, Zou W (2019) Deep-learning-powered photonic analog-to-digital conversion. Light Sci Appl 8(1), Article ID 66
16. Tu D, Huang X, Liu Y, Yu Z, Li Z (2023) Photonic sampled and quantized analog-to-digital converters on thin-film lithium niobate platform. Opt Express 31(2):1931–1942
17. Byun H, Pudo D, Frolov S, Hanjani A, Shmulovich J, Ippen EP, Kärtner FX (2009) Integrated low-jitter 400-MHz femtosecond waveguide laser. IEEE Photonics Technol Lett 21(12):763–765
18. Shtyrkova K, Callahan PT, Li N, Magden ES, Ruocco A, Vermeulen D, Kärtner FX, Watts MR, Ippen EP (2019) Integrated CMOS-compatible Q-switched mode-locked lasers at 1900nm with an on-chip artificial saturable absorber. Opt Express 27(3):3542–3556
19. Singh N, Ippen E, Kärtner FX (2020) Towards CW modelocked laser on chip-a large mode area and NLI for stretched pulse mode locking. Opt Express 28(15):22562–22579
20. Singh N, Lorenzen J, Sinobad M, Wang K, Liapis AC, Frankis HC, Haugg S, Francis H, Carreira J, Geiselmann M, Gaafar MA, Herr T, Bradley JDB, Sun Z, Garcia-Blanco SM, Kärtner FX (2024) Silicon photonics-based high-energy passively Q-switched laser. Nat Photonics 18:485–491
21. Tabatabaei Mashayekh A, Klos T, Geuzebroek D, Klein E, Veenstra T, Büscher M, Merget F, Leisching P, Witzens J (2021) Silicon nitride PIC-based multi-color laser engines for life science applications. Opt Express 29(6):8635–8653
22. Singh N, Lorenzen J, Wang K, Gaafar MA, Sinobad M, Francis H, Edelmann M, Geiselmann M, Herr T, Garcia-Blanco SM, Kärtner FX (2025a) Watt-class silicon photonics-based optical high-power amplifier. Nat Photonics 19(3):307–314
23. Richardson DJ, Nilsson J, Clarkson WA (2010) High power fiber lasers: current status and future perspectives. J Opt Soc Am B 27(11):B63–B92
24. Singh N, Lorenzen J, Wang K, Gaafar MA, Sinobad M, Francis H, Edelmann M, Geiselmann M, Herr T, Garcia-Blanco SM et al (2025b) Watt-class silicon photonics-based optical high-power amplifier. Nat Photonics 19:307–314
25. Singh N, Lorenzen J, Kilinc M, Wang K, Sinobad M, Francis H, Carreira J, Geiselmann M, Demirbas U, Pergament M et al (2025c) Sub-2W tunable laser based on silicon photonics power amplifier. Light Sci Appl 14(1), Article ID 18
26. Becker PM, Olsson AA, Simpson JR (1999) Erbium-doped fiber amplifiers: fundamentals and technology. Elsevier
27. Sinobad M, Molteni F, Gehrmann P, Lorenzen J, Hammer DN, Gaafar MA, Herr T, Singh N, Kärtner FX (2025) Dispersion compensating silicon nitride waveguide Bragg gratings for ultrashort pulse compression. In: Laser science to photonic applications (CLEO)
28. Zazzi A, Müller J, Gudyriev S, Marin-Palomo P, Fang D, Scheytt JC, Koos C, Witzens J (2020) Fundamental limitations of spectrally-sliced optically enabled data converters arising from MLL timing jitter. Opt Express 28(13):18790–18813
29. Rahman MA, Valdez F, Mere V, de Beeck CO, Wuytens P, Mookherjea S (2025) High-performance hybrid lithium niobate electro-optic modulators integrated with low-loss silicon nitride waveguides on a wafer-scale silicon photonics platform. arXiv:2504.00311
30. Rahim A, Goyvaerts J, Szelag B et al (2019) Open-access silicon photonics platforms in Europe. IEEE J Select Top Quantum Electron 25, Article ID 8200818
31. Young IT, van Vliet LJ (1995) Recursive implementation of the Gaussian filter. Signal Process 44:139–151

32. Rosales R, Murdoch SG, Watts RT et al (2012) High performance mode locking characteristics of single section quantum dash lasers. Opt Express 20:8649–8657
33. Moscoso-Mártir A, Tabatabaei-Mashayekh A, Müller J et al (2018) 8-channel WDM silicon photonics transceiver with SOA and semiconductor mode-locked laser. Opt Express 26:25446–25459
34. Pfeifle J, Brasch V, Lauermann M, Yu Y, Wegner D, Herr T, Hartinger K, Schindler P, Li J, Hillerkuss D, Schmogrow R, Weimann C, Holzwarth R, Freude W, Leuthold J, Kippenberg TJ, Koos C (2014) Coherent terabit communications with microresonator kerr frequency combs. Nat Photonics 8(5):375–380
35. Trocha P, Karpov M, Ganin D, Pfeiffer MHP, Kordts A, Wolf S, Krockenberger J, Marin-Palomo P, Weimann C, Randel S, Freude W, Kippenberg TJ, Koos C (2018) Ultrafast optical ranging using microresonator soliton frequency combs. Science 359(6358):887–891
36. Drayss D, Fang D, Sherifaj A, Peng H, Füllner C, Henauer T, Lihachev G, Harter T, Freude W, Randel S, Kippenberg TJ, Zwick T, Koos C (2025) Optical arbitrary waveform generation (oawg) using actively phase-stabilized spectral stitching. arXiv:2412.09580
37. Zazzi A, Witzens J (2024) Chip-to-chip orthogonal delay-division multiplexed network with optically enabled equalization. J Lightwave Technol 42(22):7937–7953
38. Feldmann J, Youngblood N, Karpov M, Gehring H, Li X, Stappers M, Le Gallo M, Fu X, Lukashchuk A, Raja AS, Liu J, Wright CD, Sebastian A, Kippenberg TJ, PWHP, Bhaskaran H (2021) Parallel convolutional processing using an integrated photonic tensor core. Nature 589:52–58
39. Kippenberg TJ, Gaeta AL, Lipson M, Gorodetsky ML (2018) Dissipative Kerr solitons in optical microresonators. Science 361(6402), Article ID eaan8083
40. Haboucha A, Zhang W, Li T, Lours M, Luiten AN, Le Coq Y, Santarelli G (2011) Optical-fiber pulse rate multiplier for ultralow phase-noise signal generation. Opt Lett 36(18):3654–3656
41. Kageyama T, Hasegawa T (2022) Mode spacing multiplication of optical frequency combs without power loss. Opt Express 30:19090–19099
42. Romero-García S, Moscoso-Mártir A, Müller J, Shen B, Merget F, Witzens J (2018) Wideband multi-stage CROW filters with relaxed fabrication tolerances. Opt Express 26(4):4723–4737
43. Nakajima Y, Nishiyam A, Yoshida S, Hariki T, Minoshima K (2017) Mode-filtering of a fiber-based optical frequency comb with long-fiber-based ring resonator for repetition rate multiplication. In: 2017 conference on lasers and electro-optics pacific rim (CLEO-PR)
44. Bogaerts W, De Heyn P, Van Vaerenbergh T, De Vos K, Kumar Selvaraja S, Claes T, Dumon P, Bienstman P, Van Thourhout D, Baets R (2012) Silicon microring resonators. Laser Photonics Rev 6(1):47–73
45. Katti R, Prince S (2018) Analysis of serial and parallel cascaded microring resonators: an FDTD approach. Optik 152:36–48
46. Cui S, Yu Y, Cao K, Pan Z, Gao X, Zhang X (2024) Integrated waveguide coupled ultralow-loss multimode waveguides based on silicon nitride resonators. Opt Express 32:2179–2187

Open Access This chapter is licensed under the terms of the Creative Commons Attribution 4.0 International License (http://creativecommons.org/licenses/by/4.0/), which permits use, sharing, adaptation, distribution and reproduction in any medium or format, as long as you give appropriate credit to the original author(s) and the source, provide a link to the Creative Commons license and indicate if changes were made.

The images or other third party material in this chapter are included in the chapter's Creative Commons license, unless indicated otherwise in a credit line to the material. If material is not included in the chapter's Creative Commons license and your intended use is not permitted by statutory regulation or exceeds the permitted use, you will need to obtain permission directly from the copyright holder.

Stitched-Spectrum Nonlinear Frequency Division Multiplexed Transmission Systems Using Photonic Integration

Stephan Pachnicke, Olaf Schulz, Alvaro Moscoso-Mártir and Jeremy Witzens

Abstract

We present the concept and design of a nonlinear frequency division multiplexed transmission system in which a broadband nonlinear spectrum is stitched together with help of a photonic integrated circuit. Four independently modulated WDM channels are linearly multiplexed together with trapezoidal filtering and partially overlapping spectra, approximating a seamless wideband nonlinear spectrum in which guard bands are avoided. We show that this approach mitigates the problem of nonlinear interaction between the channels, enabling more efficient use of the available spectrum. The system is highly scalable and allows for add-drop multiplexing using optic-electronic-optic signal processing. Different forms of nonlinear spectrum modulation are discussed and the system is simulated to establish its performance limits. A data rate of 400 Gbps is shown to be supported over 800 km with single-polarization modulation and a spectral efficiency of 4 b/s/Hz, with a BER staying below the SD-FEC limit for every subcarrier.

S. Pachnicke (✉) · O. Schulz
Chair of Communications, Kiel University, Kiel, Germany
e-mail: stephan.pachnicke@tf.uni-kiel.de

A. Moscoso-Mártir · J. Witzens
Institute of Integrated Photonics, RWTH Aachen University, Aachen, Germany

Acronyms

ADC	Analog-to-Digital Converter
AMF	Advanced Micro Foundry
AMP	Amplifier
APC	Active Phase Control
APFSK	Amplitude Phase Frequency Shift Keying
APSK	Amplitude Phase Shift Keying
ASE	Amplified Spontaneous Emission
ASK	Amplitude Shift Keying
AWG	Arbitrary Waveform Generator
BER	Bit Error Ratio
BPD	Balanced Photodiodes
BW	Bandwidth
CROW	Coupled (Ring-)Resonator Optical Waveguide
DAC	Digital-to-Analog Converter
DC	Direct Current
DCS	Directional Coupler Splitter
DSP	Digital Signal Processing
DUV	Deep Ultra-Violet
E-O	Electro-Optic
EDFA	Erbium Doped Fiber Amplifier
EV	Eigenvalue
EVM	Error Vector Magnitude
FBMC	Filter Bank Multi-Carriers
FSK	Frequency Shift Keying
FSR	Free Spectral Range
GC	Grating Coupler
GCPW	Grounded Co-Planar Waveguides
HD-FEC	Hard Decision Forward Error Correction
I/O	Input/Output
IL	Insertion Loss
INFT	Inverse Nonlinear Fourier Transform
IQ	Inphase-Quadrature
IXT	Inter-Channel Crosstalk
LO	Local Oscillator
MMI	Multi-Mode Interferometer
MMSE	Minimum Mean Square Error
MZM	Mach-Zehnder Modulator
NFDM	Nonlinear Frequency Division Multiplexing
NFT	Nonlinear Fourier Transform

NLSE	Nonlinear Schrödinger Equation
NN	Neural Network
OADM	Optical Add-Drop Multiplexer
ODL	Optical Delay Line
OEO	Optical-Electrical-Optical
OFDM	Orthogonal Frequency Division Multiplexing
OH	Optical Hybrid
OSA	Optical Spectrum Analyzer
PCB	Printed Circuit Board
PD	Photodiode
PIC	Photonic Integrated Circuit
PN	Phase Noise
PS	Phase Shifter
PSD	Power Spectral Density
PSK	Phase Shift Keying
QAM	Quadrature Amplitude Modulation
RBW	Resolution Bandwidth
RF	Radio-Frequency
RIN	Relative Intensity Noise
SC	Subcarriers
SD-FEC	Soft Decision Forward Error Correction
SDM	Space-Division Multiplexing
SE	Spectral Efficiency
SiP	System-in-Package
SOI	Silicon-on-Insulator
SSMF	Standard Single-Mode Fiber
SVM	Support Vector Machine
TEC	Thermoelectric Cooler
WDM	Wavelength Division Multiplexing

6.1 Introduction

Fiber-optic communication systems are the backbone of today's global digital infrastructure. Fulfilling the ever-increasing demand for global data traffic is a challenge because the Kerr effect limits the achievement of higher spectral efficiency (SE) through increased optical power [1]. Approaches such as space-division multiplexing (SDM) circumvent this limitation by using multiple fiber cores or modes to distribute the power across multiple paths [2]. This does, however, require a complete overhaul of the optical fiber infrastructure, making data-processing based schemes increasing the SE in existing fiber infrastructure of high interest. One promising approach is the use of the nonlinear Fourier transform (NFT).

Its spectrum evolves linearly along the optical fiber channel and reduces the required equalization in the receiver to a simple linear phase rotation [3]. This opens the possibility to improve the nonlinear Shannon limit faced by current systems, in which fiber nonlinearities create an additional limit on the high signal power side as illustrated in Fig. 6.1. While the achievable capacity is fundamentally limited by noise and by the Shannon limit on the low signal power side, the nonlinearities affecting signal transmission at high power may become less limiting as approaches applied to compensate them improve.

In contrast to the well-known Fourier transform, the NFT results in a continuous and a discrete spectrum. The discrete part describes the solitonic portion of the spectrum, which can be used by itself to transmit data [4–8]. The continuous part consists in dispersive waves and is similar to the well-known linear Fourier spectrum to which it reduces in the limit of low optical power. Similarly, it can be modulated using orthogonal subcarriers, according to a modulation scheme called nonlinear frequency division multiplexing (NFDM) [9] which has been intensely investigated in the last years [10–16]. Both parts of the spectrum can be jointly modulated with the objective to improve the SE [17–20].

Figure 6.2 shows the generic architecture of a single-channel long-haul NFDM transmission system. The hardware components are the same as in a conventional coherent system, so that the only difference is in the signal processing part. In the transmitter, modulated nonlinear spectra are generated in the form of overlapping and mutually orthogonal sinc functions, transformed into the time domain using the inverse (I)NFT, and processed further like any other time-domain signal. After applying the NFT in the receiver, the linear phase rotation is applied before the data is extracted. Other basic steps such as timing-synchronization, carrier frequency offset compensation and phase noise compensation are not shown in this diagram, but are also necessary as in any real system.

Despite the intensity of previous research efforts, there are still significant challenges that need to be overcome in order to outperform conventional systems in realistic scenarios. Lumped amplification with erbium doped fiber amplifiers (EDFAs) in existing infrastructure and nonlinear noise-noise interaction [21] in particular break the regularities of the NFT and present significant obstacles towards achieving higher SE. This has motivated further research into more advanced data-processing methods such as neural networks [22–26], probabilistic shaping [27, 28], or the use of different subcarrier waveforms [29–31]. State-of-the-art NFT-based transmission systems are able to achieve data rates of up to 20.48

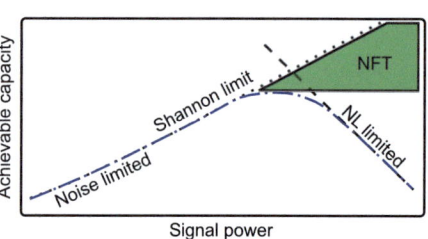

Fig. 6.1 Capacity limit of optical fiber communication systems

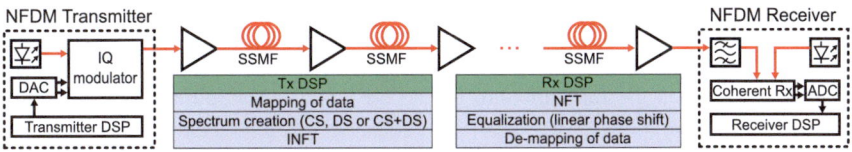

Fig. 6.2 General structure of a single-channel long-haul NFDM transmission system. CS: continuous spectrum, DS: discrete spectrum

Tb/s using wavelength division multiplexing (WDM) [32], 1.004 Tb/s with only a single channel [33], and a spectral efficiency of 7.72 b/s/Hz has been demonstrated for a single polarization system using filter bank multi-carriers (FBMC) [31]. However, most of the work has focused on single-channel transmission with limited bandwidth, and scaling over large spectra, such as the entire C-band, remains an important challenge. The achievable bandwidth is generally limited by the finite modulation speed of electro-optic (E-O) components and by the high computational complexity of the I/NFT for signals with a high number of samples [34, 35]. The combination of NFDM with WDM is not a straightforward solution to this issue because there are no nonlinear multiplexers available and linear multiplexing requires guard bands between the channels to avoid nonlinear inter-channel crosstalk (IXT) [32, 36, 37]. E-O bandwidth limitations can be overcome by stitching together spectral slices [38, 39], as already applied to NFDM systems [40], but this does not solve the problem of an impractically high computational complexity. Additionally, solutions have to be found to enable add-drop multiplexing in the framework of NFDM.

In the following, we will present a transmitter concept for NFDM WDM transmission using a photonic integrated circuit (PIC) that overcomes the bandwidth and computational limitations outlined in the previous paragraph. Individually generated and spectrally overlapping nonlinear spectra are combined with a weighted average in the linear domain so as to mitigate the IXT between them [41–43]. Guard bands are no longer required so that all available bandwidth can be used for data transmission and the SE is increased. In additon, the computational complexity of the NFT is kept low, since digital signal processing (DSP) is applied separately to the individual channels. Since spectral stitching requires the relative phases of the different channels to remain stable, implementation of the transmitter with a PIC is highly advantageous over a hybrid fiber-optic system. Our solution is highly scalable and is only limited by physical layouting constraints and heat sinking of the PIC. This system architecture concept is also flexible as it can be extended to allow for add-drop multiplexing using optic-electronic-optic circuits [44–48].

The chapter is structured as follows: Section 6.2 gives a basic overview over the NFT and the modulation of the two nonlinear spectra. Next, we describe the challenges when building a WDM NFDM transmission system and present our transmitter concept in Sect. 6.3. The details for its practical implementation are described in Sect. 6.4. After that, we validate the concept through realistic simulations in Sect. 6.5 and extend it to more flexible systems with nonlinear add-drop multiplexing in Sect. 6.6. Section 6.7 concludes and gives an outlook for further work.

6.2 Nonlinear Fourier Transform Basics

This section describes the fundamentals of the NFT and the modulation of data onto the nonlinear spectra.

6.2.1 Mathematical Background

The propagation of light in an optical fiber can be described with a partial differential equation, the nonlinear Schrödinger equation (NLSE). In the framework of the NFT, the NLSE is used in a normalized form that also neglects losses and noise [3, 9]:

$$j\frac{\partial q(t,z)}{\partial z} = \frac{\partial^2 q(t,z)}{\partial t^2} + 2|q(t,z)|^2 q(t,z), \tag{6.1}$$

where $q(t,z)$ is the complex-valued envelope of the electric field, z is the propagation distance and t is the time. Following the normalization scheme used in [9], the free parameter T_0 is set to 1 ns in our case. The so-called Zakharov-Shabat problem [49] is derived from the NLSE and is given by

$$\frac{\partial v}{\partial t} = \begin{pmatrix} -j\lambda & q(t) \\ -q^*(t) & j\lambda \end{pmatrix} v, \quad v(T_1, \lambda) = \begin{pmatrix} 1 \\ 0 \end{pmatrix} e^{-j\lambda T_1}. \tag{6.2}$$

For a given nonlinear frequency λ, Eq. (6.2) can be solved numerically in an interval $[T_1, T_2]$ containing the analyzed optical pulse to find the Jost scattering coefficients $a(\lambda)$ and $b(\lambda)$ defined as

$$\begin{aligned} a(\lambda) &= v_1(T_2, \lambda) e^{j\lambda T_2}, \\ b(\lambda) &= v_2(T_2, \lambda) e^{-j\lambda T_2}. \end{aligned} \tag{6.3}$$

The nonlinear spectrum consists of the continuous and discrete spectra. The continuous spectrum is defined for $\lambda \in \mathbb{R}$ and is given by

$$q_c(\lambda) = \frac{b(\lambda)}{a(\lambda)}. \tag{6.4}$$

The discrete spectrum is defined by the zeros λ_j of $a(\lambda)$ in the upper half complex plane:

$$q_d(\lambda_j) = \frac{b(\lambda_j)}{da(\lambda)/d\lambda|_{\lambda=\lambda_j}}, \quad j = 1, 2, \ldots, M. \tag{6.5}$$

As mentioned in the introduction, these nonlinear spectra evolve with linear phase increments in the fiber, which is a key property of the NFT. The evolution of the NFT-spectrum is given by

$$q(\lambda, z) = e^{-4j\lambda^2 z} q(\lambda, 0). \tag{6.6}$$

6.2.2 Continuous Spectrum Modulation

The continuous part, as the counterpart to the linear Fourier spectrum, can be modulated by summing up orthogonal subcarriers, similar to orthogonal frequency division multiplexing (OFDM). Although various subcarrier forms are possible, sinc-carriers are commonly used:

$$q_c(\lambda) = A \sum_{k=-(N-1)/2}^{(N-1)/2} C_k \operatorname{sinc}\left(\frac{T_c}{\pi}\lambda + k\right), \tag{6.7}$$

where N is the number of subcarriers, C_k are the symbols drawn from an arbitrary quadrature amplitude modulated (QAM) constellation diagram, T_c is the useful block duration and A is the power control parameter. T_c results from the division of N by the channel bandwidth and is 4 ns in the following, unless otherwise specified.

The time-domain signal of the q_c-modulated spectrum resulting from the INFT implemented with a suitable numerical algorithm [3, 50] is, however, not bounded in time and consists in a pulse that has a large tail [51] that degrades the performance or the spectral efficiency of the system, since either the signal has to be truncated or the time window in which individual data-blocks are launched has to be increased. The solution is to modulate the b-coefficients instead of q_c. This modulation scheme, referred to as b-modulation, leads to time limited signals, but introduces the drawback that the launch power is now limited [52]. To ensure that this condition is satisfied as well as to increase the overall launch power within that achievable envelope, an exponential scaling can be applied to the initial spectrum $u(\lambda)$ modulated according to Eq. (6.7) using the Γ-transform as [53, 54]

$$b(\lambda) = \sqrt{1 - e^{-|u(\lambda)|^2}} e^{j\angle u(\lambda)}. \tag{6.8}$$

The differences in power are clearly visible in Fig. 6.3, in which the continuous spectra resulting from modulation with the three methods are shown.

As a final step, the phase compensation necessary according to Eq. (6.6) can already be partially implemented at the transmitter. By applying half at the transmitter (pre-compensation) and half at the receiver (post-compensation), the time-domain guard interval between the blocks that prevents inter-block interference can be halved [55]. The final

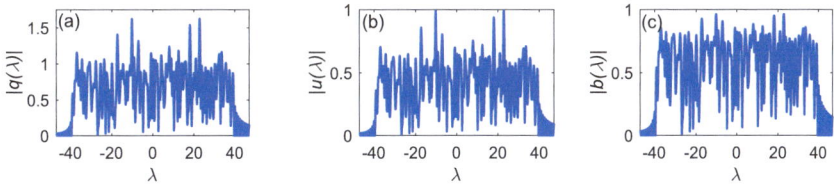

Fig. 6.3 Nonlinear continuous spectra resulting from the different methods of modulation. **a** q-modulation, **b** b-modulation, **c** b-modulation with Γ-transform

modulated continuous spectrum is therefore generated as

$$b(\lambda) = Ae^{2j\lambda^2 z}\Gamma_b \left(\sum_{k=-(N-1)/2}^{(N-1)/2} C_k \mathrm{sinc}\left(\frac{T_c}{\pi}\lambda + k\right) \right). \quad (6.9)$$

6.2.3 Discrete Spectrum Modulation

The eigenvalues forming the discrete spectrum, each of which would correspond, on its own, to a first-order soliton, are located in the upper half complex plane and have complex-valued spectral amplitudes, so that in total four degrees of freedom exist, which can be independently modulated.

$\Re(\lambda_j)$ is linked to the center frequency of the soliton and $\Im(\lambda_j)$ defines the amplitude and therefore also the width of the soliton pulse in the time domain. It is common to modulate $b(\lambda_j)$ instead of $q_d(\lambda_j)$ as it is more robust to noise [5]. Both, the phase and the absolute value of $b(\lambda_j)$ can be independently modulated using an amplitude phase shift keyed (APSK) modulation format. The absolute value of $b(\lambda_j)$ is linked to the timing of the soliton according to

$$|b(\lambda_j)| = e^{2\Im(\lambda_j)\Delta T}, \quad (6.10)$$

where ΔT is the normalized time shift. The amplitude modulation applied to $|b(\lambda_j)|$ is limited by the soliton pulse falling within and remaining in the time window allocated to its transmission block, especially when combined with the continuous spectrum. Figure 6.4 shows the discrete spectrum with one eigenvalue and the resulting time-domain signal. It can be seen that the normalized amplitude in the time domain corresponds to two times the imaginary part of the eigenvalue [3] and that the soliton is shifted in time by the amount determined by the discrete spectrum coefficient.

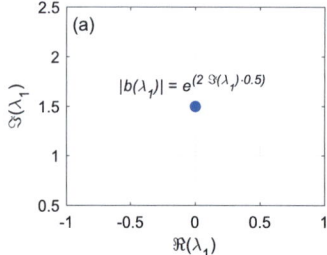
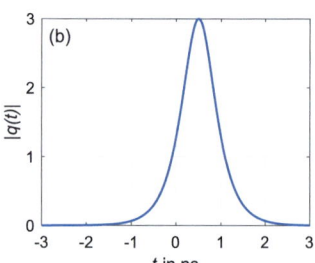

Fig. 6.4 a Modulated discrete spectrum with one eigenvalue. b Resulting time-domain signal

6.2.4 Full Spectrum Modulation

If both parts of the nonlinear spectrum are used for the transmission, they must be fed together into the INFT algorithm, since an independent calculation and subsequent addition does not lead to the correct result by nature of the nonlinear interactions. Figure 6.5 shows a fully modulated nonlinear spectrum, the resulting time-domain signal, and its linear spectrum. The modulated spectrum has a bandwidth of 25 GHz, consists of 100 subcarriers and two eigenvalues positioned 1 ns to the left and to the right of the center of the overall pulse, as can be seen in Fig. 6.5b. The solitons have a frequency offset of −3 and +3 GHz, which is also apparent in Fig. 6.5c.

6.3 Transmitter Concept

A single-channel NFDM transmission can be directly deployed because the NFT accounts for all Kerr-nonlinearities occurring in the fiber. However, single-channel transmission is severely limited by the finite bandwidth of the E-O hardware, as for any transmitter system. Furthermore, the computational complexity of fast (I)NFTs scales as $O(n \log^2 n)$, where n is the number of samples [56], which further constrains practical transmission systems, so that multiple channels (WDM) have to be used.

The block diagram of an NFDM WDM transmission system is shown in Fig. 6.6. The signals of each WDM channel are first generated independently, after which they are multiplexed together, before being sent down the transmission link. At the receiver, the signal

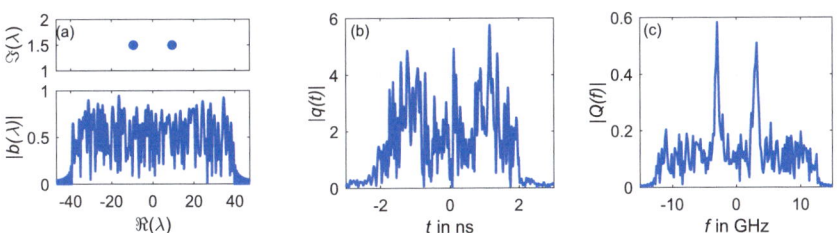

Fig. 6.5 a Modulated continuous and discrete spectrum. **b** Resulting signal in the time domain. **c** Corresponding linear spectrum

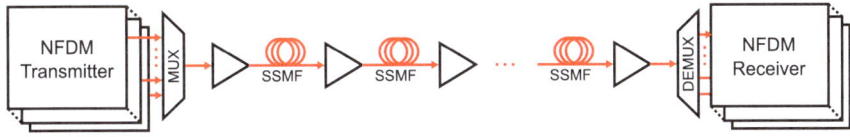

Fig. 6.6 Block diagram of an NFDM WDM transmission system

Fig. 6.7 BER versus spectral guard band width for a two-channel NFDM WDM system with 100 subcarriers each

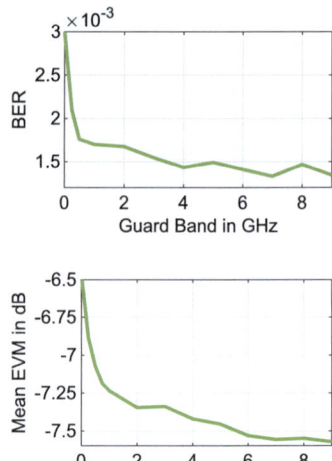

Fig. 6.8 Mean EVM of the subcarriers at the edge of the spectrum versus spectral guard band width

is demuliplexed and each channel is evaluated separately. Since nonlinear multiplexers do not exist, the channels have to be linearly multiplexed, which introduces nonlinear crosstalk between them that is not handled by the NFDM framework, as the summation of independent INFTs over the channels is not the same as the INFT being applied to all the channels jointly:

$$\sum_{k=1}^{N_c} \text{INFT}(q_k(\lambda)) \neq \text{INFT}\left(\sum_{k=1}^{N_c} q_k(\lambda)\right), \quad (6.11)$$

where N_c is the number of WDM channels. The most common approach to avoid IXT is to use frequency guard bands between the channels, which reduces the SE, as these parts of the spectrum can no longer be used for data transmission [32, 57].

To exemplarily demonstrate the effect of IXT in an NFDM WDM transmission system, Fig. 6.7 shows the overall bit error ratio (BER) for two channels transmitted jointly with different guard bands. The number of subchannels per channel is set to 100, as previously, and only the continuous spectrum is modulated. It can be seen that without a guard band the performance drops significantly. The underlying penalty is visualized in Fig. 6.8, in which the mean error vector magnitude (EVM) of the subcarriers at the edges of the two channels is shown. Even if the performance is already significantly improved with a guard band of 1 GHz, it shows that a nonlinear IXT penalty remains up to a guard band of 6 GHz, after which the curve reaches a plateau determined by EDFA noise only.

From these simulations, it is clear that there is a significant performance degradation in the subcarriers situated at the edges of the NFDM channels, if these are linearly multiplexed with insufficient guard bands. Hence, it is crucial to find new ways to linearly multiplex the

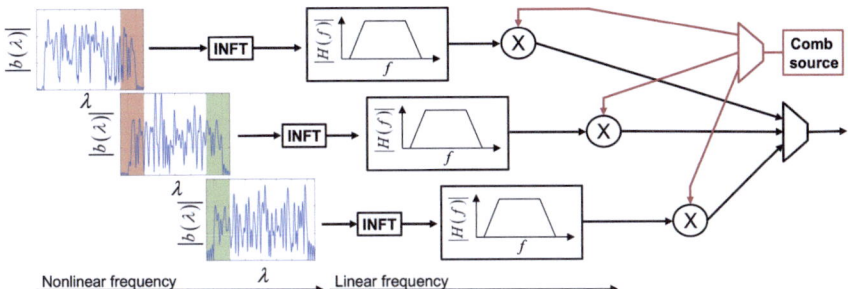

Fig. 6.9 Diagram of the signal generation flow. f refers to the linear frequency and $H(f)$ to the filter transfer functions [41]

NFDM channels that can mitigate IXT even when channels are independently processed by the INFT.

To that effect, we propose a solution whose key concept is to increase the bandwidth of the independently generated NFDM channels in the nonlinear Fourier domain so as to create a spectral overlap between neighboring channels. Subcarriers placed in the overlap regions are loaded with the same data and are processed twice, once for each channel they belong to. Sinc-shaped nonlinear spectra with central lobes in the color-coded overlap regions in Fig. 6.9 are thus identical. Note that there are still differences in the overall spectra shown in these regions due to the adjacent subcarriers from outside the overlap region, as their sidelobes also extend to it and they are different for the two channels.

After mapping to the linear domain with the INFT, linear filters with a trapezoidal-shaped amplitude transfer function are applied. The ramp-shaped filter edges located in the spectral overlap regions sum up to one and effectively define a weighted average that assigns, for each subchannel, a stronger weight to the INFT of the channel that includes the highest number of neighboring subchannels. This provides a gradual transition from the results of one INFT to the other, while accounting for the nonlinear interaction of the subcarriers with their nearest neighbors from within the overlap region and from the neighboring channels.

Finally, the resulting signals are modulated and linearly combined in the optical domain.

Even though this is not strictly the same as applying an INFT over the entire spectrum and some level of IXT thus remains, this approach provides a suitable approximation as a consequence of the Kerr-mediated nonlinear interaction strength decaying with the center frequency difference between the subcarriers, which in turn results from fiber dispersion [57]. The weighted average defined by the shape of the trapezoidal filters is biased towards taking into account the most relevant nonlinear interactions with the closest subcarriers throughout the overall spectrum. The two spectra are equally weighted for the subcarrier in the center of the overlap region, which is also the subcarrier with which their nonlinear interaction is equally strong. At the edges of the overlap regions, the channel closest to the subcarrier, for which the ramp approaches one, is primarily taken into account.

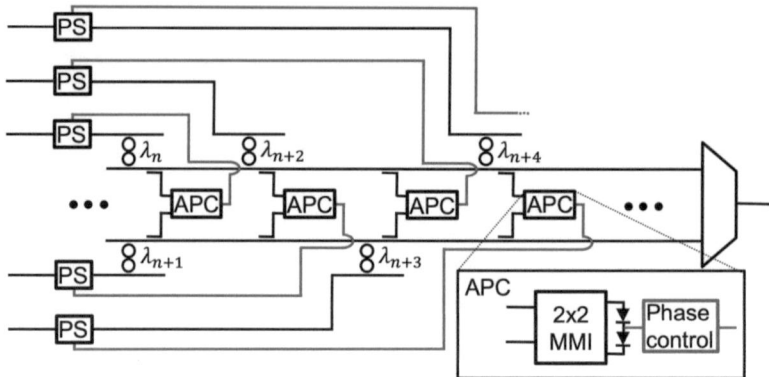

Fig. 6.10 Block diagram of the cascaded multiplexing concept in which active phase controllers are used to maintain the phase coherence between adjacent channels [41]

This novel concept, first proposed in [41, 43], alleviates some of the constraints limiting current NFDM systems: (i) E-O modulator bandwidth requirements are reduced by splitting the spectrum over multiple channels, (ii) the overall computational complexity is reduced by applying individual INFTs to individual channels, and (iii) nonlinear interactions are accounted for in a seamlessly stitched spectrum.

One of the challenges of this approach is that the individual NFDM channels have to be coherently added up in the optical domain so that the signal does not get destroyed in the overlap regions. This can be achieved using phase coherent carriers, such as generated by combs as also used in more conventional WDM systems [58], to generate the individual optical channels and by dynamically correcting residual phase errors before multiplexing. By synchronizing the repetition rate of the comb with the clock used to generate the electronic signals and balancing the optical paths implemented on the PIC, long-term stability of the system can be obtained [38]. The phase errors introduced by the different optical paths inside the PIC prior to multiplexing of the channels can be corrected using active phase control (APC) as demonstrated in [39]. The main drawback of the cascaded solution presented there is that it scales poorly with the number of channels due to the $\sim$6 dB losses introduced in each stage.

To make the system more scalable, we build on the architecture introduced in [8] in which two bus waveguides are used to aggregate the channels with even and odd indices, respectively, that are subsequently combined with a 2-by-1 combiner that creates a one-time penalty of 3 dB, see Fig. 6.10. This way, channels added to the buses do not spectrally overlap with any other channel on that bus, so that wideband coupled (ring-)resonator optical waveguide (CROW) add-drop multiplexers (OADMs) with an optical bandwidth exceeding the channel width can be used without introducing any spectral clipping.

In order to establish phase coherence between the channels, a small amount of light (1%) is tapped off from the two buses every time a channel has been added to one of them and is fed to a 2-by-2 multi-mode interferometer (MMI) followed by a balanced photodetector pair. A feedback loop sets an upstream phase shifter (PS) that adjusts the phase of the added channel until the power levels recorded from the two photodiodes are equal, which is the interference condition corresponding to the added channel being in phase, inside the overlap region, with the adjacent preexisting one already present in the opposite bus waveguide. The PSs are recursively set until the entire spectrum is coherently combined. This concept can be scaled up to multiplex a large number of channels and in its current implementation is primarily limited by the number of modulators that can be fit in the chip and by its overall thermal load, see Sect. 6.4.

6.3.1 Receiver Architecture

Standard receivers can be used to recover the individual channels, which is beneficial to keep the hardware complexity as low as possible. In contrast to a spectrum with guard bands, where the channels can be filtered out one by one from the incoming signal, this is not directly possible with the seamless spectrum generated by our transmitter, since subcarriers would be partially dropped at the edges of the filters. A possible solution to this problem would be to digitally stitch the spectral parts back together [38], but this would require more complex hardware and sophisticated DSP. A simpler method, which is implemented in the simulation setup, is to split the signal with a 3-dB splitter. All even channels are then filtered out from one output and all odd channels from the other one. As for the transmitter, the resulting 3-dB loss also occurs only once, regardless of the number of channels. This architecture is similar to the two bus architecture used in the transmitter, however, phase coherence is not required here, as the spectra are not stitched back together. Rather, each channel is dropped with a sufficiently wideband filter such that all its subchannels are surrounded by a sufficient amount of modulated spectrum.

6.3.2 Transmission with Continuous Spectrum Modulation

Figure 6.11 shows the simulation setup that has been used to validate the system concept. The overall system is again taken to have a bandwidth (BW) of 100 GHz (centered at

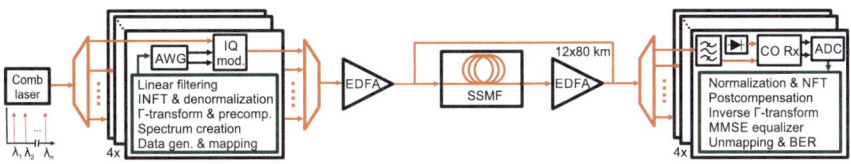

Fig. 6.11 Block diagram of the transmission system simulated for system concept validation

1550 nm) split over 4 channels, each having a BW in excess of 25 GHz to account for the required overlap. After the data is generated and mapped to either 16-QAM or 32-QAM symbols, the b-modulated spectrum is created for each channel. After that, the Γ-transform and dispersion pre-compensation are applied, as explained in Sect. 6.2.2. Finally, the INFT is calculated, trapezoidal-like linear filtering is applied to each channel and the electrical drive-signals needed for modulation are generated with an arbitrary waveform generator (AWG) with an 88 GSa/s sampling rate. After optical modulation, all channels are coherently multiplexed and sent over 12 spans of 80 km standard single-mode fiber (SSMF) with the EDFAs set to an output power of -1 dBm. On the receiver side, channels are de-multiplexed, filtered and demodulated using a coherent receiver and an 80 GSa/s analog-to-digital converter (ADC). The NFT is calculated, remaining dispersion compensation is applied and the inverse Γ-transform is computed. Before BER calculation, a simple minimum mean square error (MMSE) equalizer is applied. The fiber is modeled as having an attenuation of 0.2 dB/km, a mean dispersion β_2 of -21.68×10^{-27} s^2/m ($D = 17$ ps/nm/km), a nonlinearity coefficient γ of 1.3 W^{-1}km^{-1}, and the EDFAs have a noise figure of 5 dB. Other than the noise figure of the EDFAs inside the fiber optic link and the sampling rates of the AWG and ADC used in the transmitter and receiver systems, no other non-idealities are considered here. A more comprehensive modeling taking the insertion losses of the PIC and the non-idealities of the light source into account is reported in Sect. 6.5.

Figures 6.12 and 6.13 show the simulated BER as a function of the spectral overlap for 16- and 32-QAM transmission of 4×25 GHz effective channels with a total of 400 subcarriers and a block length of 6 ns, resulting in an SE of 2.67 b/s/Hz (16-QAM) and 3.3 b/s/Hz (32-

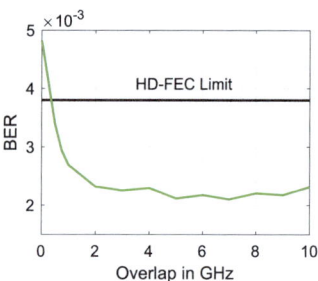

Fig. 6.12 BER versus spectral overlap for 16-QAM transmission

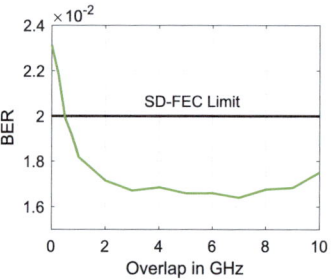

Fig. 6.13 BER versus spectral overlap for 32-QAM transmission

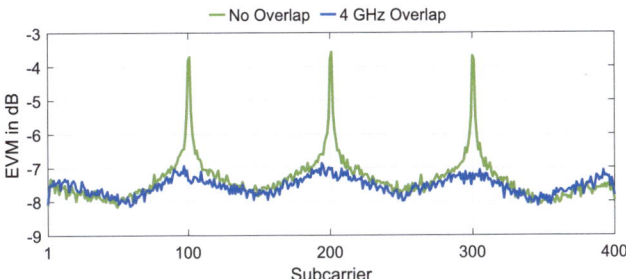

Fig. 6.14 Subcarrier EVM for 32-QAM without and with 4 GHz spectral overlap

QAM). It reveals a clear performance gain when inter-channel spectral overlap is used. Just 1 GHz overlap already reduces the BER enough to achieve error-free transmission when assuming 7% overhead hard decision forward error correction (HD-FEC) with a limit of 3.7×10^{-3} for 16-QAM or 20% overhead soft decision (SD)-FEC with a limit of 2×10^{-2} for 32-QAM. This performance can be further enhanced by using neural network equalizers [43], dual polarization (double SE), probabilistic shaping [10], or joint modulation with the discrete part of the spectrum [9]. However, Figs. 6.12 and 6.13 also show that excessively large overlaps are not beneficial, since the BER goes up again beyond ~ 7 GHz. This can be explained by considering that the linear filtering and summation in the spectral overlap regions violates the regularities of the NFT and is thus detrimental when applied where it is not needed, far from the edges of the channels.

Figure 6.14 shows the EVM for each subcarrier without and with 4 GHz spectral overlap for the 32-QAM modulation scenario. As expected, the subcarriers located at the edges of the channels are most affected, and their performance can be clearly improved by using this overlap technique at the cost of a slight increase in the required E-O BW per channel.

6.3.3 Transmission with Full Spectrum Modulation

Up to now, only the continuous part of the nonlinear spectrum has been used to benchmark the spectral overlap concept. To investigate the performance when using the full spectrum, the four-channel NFDM WDM transmission system from the previous section was simulated again with the addition of a modulated discrete spectrum. The case with a 4 GHz overlap is selected, resulting in channels with a bandwidth of 28 GHz, each containing 112 16-QAM modulated subcarriers. The overhead of the three 4 GHz overlap regions was equally split among the four channels to obtain equal channel bandwidth requirements: Subtracting the shared 3×4 GHz in the overlap regions and the corresponding 3×16 duplicate subcarriers, this can be seen to correspond to the 100 GHz, 400 subcarrier overall capacity already explored in the previous section. In addition, the discrete spectrum consists of two eigenvalues per channel with frequencies offset by -3 and 3 GHz relative to the channel center,

Fig. 6.15 BER of the continuous spectrum versus different eigenvalue amplitudes

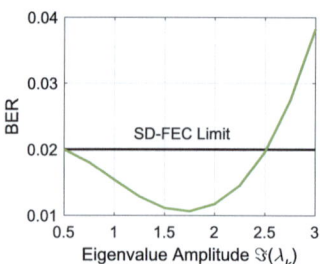

Fig. 6.16 EVM of the subcarriers in the overlap regions and around the eigenvalue frequencies versus different eigenvalue amplitudes

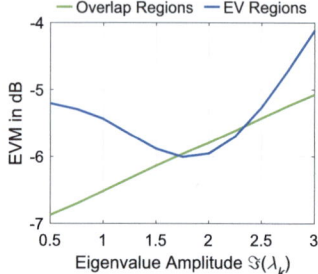

keeping them far from the edges of the channel spectrum and thus reducing spectral clipping and dispersion mediated broadening of the overall data block. Data is only modulated on the phase of $b(\lambda_j)$ using 16-PSK modulation. The spectral efficiency is therefore slightly increased compared to the previous case from 2.67 to 2.72 b/s/Hz.

The BER computed for the continuous spectrum is shown in Fig. 6.15 for different eigenvalue amplitudes. The two spectral parts of the nonlinear spectrum are strictly orthogonal to each other only in the absence of noise and losses, so that in a real link the eigenvalues and the continuous spectrum interact with each other. Since the eigenvalues contain a lot of power, they are more robust against this than the continuous spectrum. The optimal performance for the continuous spectrum is achieved for a normalized eigenvalue amplitude of 1.75, while for amplitudes higher than 2.5 and lower than 0.5 the BER increases beyond the assumed 20% SD-FEC limit of 2×10^{-2}. The BER of the discrete spectrum at the optimal point is only 8.75×10^{-4}.

For a more detailed analysis on how the eigenvalues influence the transmission quality of the continuous spectrum, Fig. 6.16 shows the EVM of the subcarriers in the overlap regions and around the eigenvalue frequencies for different eigenvalue amplitudes. Figure 6.17 shows the EVM for all the subcarriers for normalized eigenvalue amplitudes of 0.75, 1.75 and 2.75.

First, it can be seen that the EVM of the subcarriers is degraded most in the frequency regions around the eigenvalues, which has already been shown in [18]. For lower amplitudes, fewer subcarriers are affected than for higher amplitudes, but the detrimental effect on these is significantly stronger. The reason for this is revealed when looking at the linear spectrum of solitons. A high amplitude leads to a wide spectrum and thus to a distribution of the power

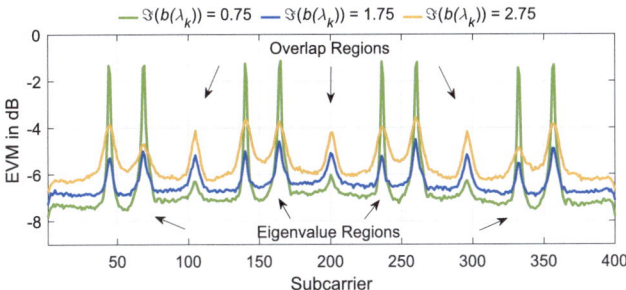

Fig. 6.17 EVM of all subcarriers versus three different eigenvalue amplitudes

over many frequencies, while at lower amplitudes the power is strongly focused in a small spectral range.

Second, the eigenvalues appear to raise the overall EVM of the subcarriers and to have a compounded effect with the EVM penalties already present in the overlap regions, that result from breaking the regularities of the NFT. With center frequencies of ± 3 GHz, the eigenvalues are positioned far away from the overlap regions, but still have a negative impact on them. The EVM increases continuously with increasing eigenvalue amplitude, which again can be explained with the broader linear spectrum of the solitons, that results in a wider part of the continuous spectrum overlaying with it in the linear domain, as well as from the increased overall power of the soliton pulses. However, the EVM in the overlap regions remains lower or very close to that of the subcarriers around the soliton frequencies, so that the latter continue to be the main limitation on the signal quality.

6.4 Transmitter System Implementation

The implementation of the transmitter presents a number of challenges. Besides the design of a suitable silicon photonics PIC, which is the central component of the transmitter, the hardware needs to support the electrical connectivity required to bring in the radio-frequency (RF) and control signals, as well as the chip-scale thermal stabilization required to maintain long-term phase coherence across the channels. A comb generator is also implemented with an E-O setup [59] outside of the chip, with a Mach-Zehnder modulator (MZM) converting a single carrier into a comb whose free spectral range (FSR) is determined by an RF signal synchronized with the clock of the remaining electronics. This guarantees long-term phase stability between the comb and the signal processing electronics, in particular the relative phase between comb lines referenced to the time base of the electronic clock. While such systems can also be implemented in silicon photonics PICs [60], the chip-scale integration of the comb generator was not essential here as all the comb lines travel down a single fiber up to the input port of the PIC, and the stability of their relative phases is thus not compromised.

In this section we are presenting the design, fabrication and characterization of the required key components, which serves as a basis for more realistic simulations in Sect. 6.5 that verify the feasibility of the system implementation. Experimental transmission characterization is ongoing and will be reported in a future publication.

6.4.1 Photonic Integrated Circuit

The system PIC is the central element of the proposed transmitter system. It implements E-O modulation and the phase-coherent optical channel multiplexing. Here, we describe the architecture of the PIC and its insertion losses and link budget. Device characteristics are based on detailed measurements of test structures located on the same chip that are also reported in the next subsection.

Figure 6.18 shows the block diagram and the layout of the system PIC. As visible in Fig. 6.18a, the PIC modulates and multiplexes up to four comb lines, but only 2 IQ-modulators are implemented, one of which modulates the even channels and the other modulates the odd channels. Since the IQ modulators are implemented as balanced nested MZMs with 2-by-2 MMIs used as splitter/combiner at the input and output of the superstructure, two wavelengths, with equal index parity, can be injected and extracted from each modulator via the pairs of complementary input-/output-ports. Since coupling to the buses occurs via 4-ring CROWs, only the intended channel is injected into the buses at a given injection point. Sharing data streams across all the even and odd channels is a common method for the characterization of broadband fiber-optic transmission in resource-constrained environments [61]. Here, it allows marginally shrinking the chip size — the 5 × 2.8 mm system chip has unused spaces on the two sides of the IQ-modulator array filled with test structures that could have been allocated to further modulators. The more important benefit is that it reduces the number of required AWG channels to four.

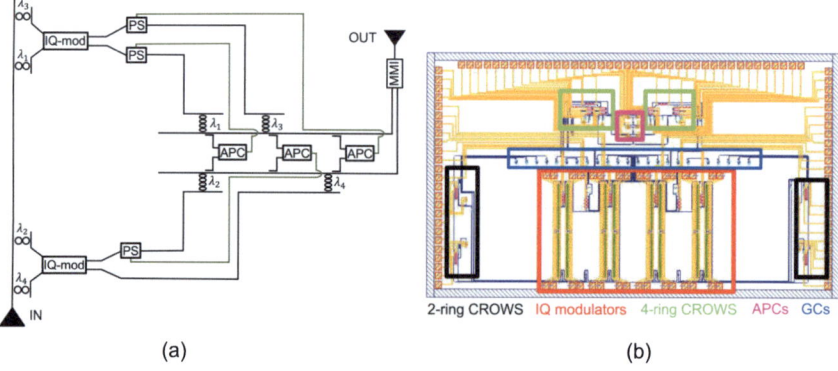

Fig. 6.18 **a** Block diagram of the PIC. **b** Layout of the PIC (5 mm × 2.8 mm) [41]

The PIC has been designed into the 220-nm device-layer-thickness silicon photonics platform of Advanced Micro Foundry (AMF), Singapore, and is compatible with deep ultra-violet (DUV) lithography at a 248-nm wavelength [62].

The comb lines are injected into the PIC using one of the grating couplers (GCs) located at the center of the chip, with an insertion loss (IL) of 3.75 dB. Once inside the PIC, up to four comb lines are filtered using 2nd-order CROW OADMs (3-dB BW = 8 GHz, IL = 3.5 dB). Up to four optical channels can be supported by tuning these in or out. After that, each carrier is independently routed to one of the two input ports of the two IQ-modulators, as explained above.

The IQ modulators are designed as nested MZMs with slotted transmission lines to improve the trade-off between 50 Ω impedance matching and RF losses, interleaved waveguide delay loops (phase recovery loops) to improve the phase matching between the RF and optical signals, and a GSSG driving scheme that suppresses RF crosstalk between complementary phase shifters when operated in a push-pull configuration [63, 64]. Moderately high doping concentrations ($\sim 2 \times 10^{18}$ cm^{-3}) are used to define the pn-junctions inside the waveguides to obtain a high phase-shifting efficiency [65] and compact rectilinear phase shifters (1100 μm). Each of the eight high-speed phase shifters are 50 Ω terminated directly on chip. The IQ-modulators have an E-O cutoff frequency above 25 GHz and an optical IL and modulation penalty of 14 dB for a 2 V_{pp} drive voltage, defined here as the ratio of the maximum optical power at their output with the power at their input, when the nested MZMs are biased at the zero-transmission point, and comprises the 3-dB losses associated to the coupler combining the I- and Q-signals.

After modulation, even and odd channels are injected into the two buses using 4th-order CROW filters (3-dB BW = 32 GHz, IL = 2 dB), architectured following the design approach in [66], so spectral crosstalk in each bus can be neglected. After each filter, a small portion of the signal (~1%) is coupled out from each bus and sent to one of the APCs. These consist of a 2-by-2 MMI, a balanced photodiode pair (BPD), and a phase shifter applied to the corresponding IQ-MZM output, as seen in Fig. 6.18.

It is essential that the optical paths of the two waveguide buses be balanced between any of the pairs of taps and the 2-by-1 combiner at the output of the PIC. To ensure this, the corresponding optical paths should also be kept as short as possible to minimize cumulative phase errors resulting from the finite coherence length of the waveguides [67]. The optical paths between the input of the PIC and the injection points into the buses are less critical, as phase errors are compensated by the APCs, however, they should also be balanced to minimize temperature dependent phase offsets, to increase the stability of the system, and to reduce the dynamic adaptability requirements of the control system. Figure 6.19 shows a detailed view of the corresponding waveguide routing, in which a close to symmetric layout has been used to improve balancing.

Finally, even and odd channels are combined from the two buses using a 2-by-1 MMI (IL = 3 dB) and the signal is coupled out to a fiber using a second GC (IL = 3.75 dB). The total losses of the PIC, including the IL from all components, the modulation penalty, and the

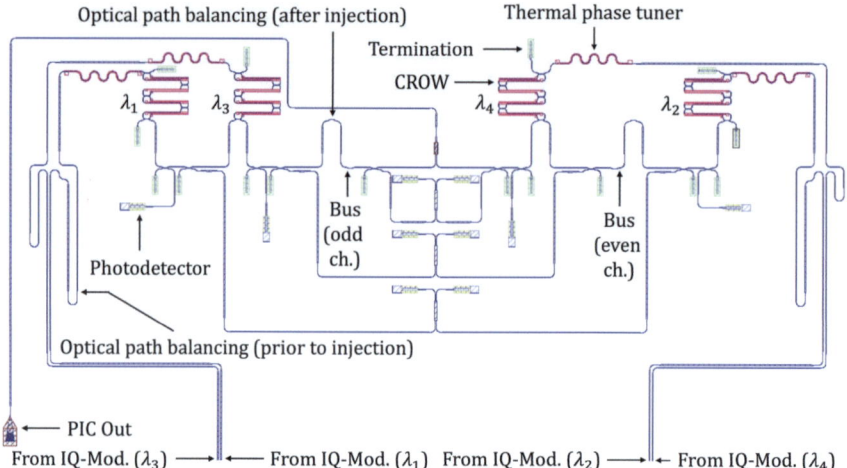

Fig. 6.19 Diagram of the waveguide routing in the multiplexer

cumulative losses of interconnect waveguides and monitor taps (IL = 3.5 dB), are 33.5 dB. This IL can be directly improved by using a higher drive voltage (e.g. 4 V_{pp} that reduces the IL and modulation penalty from 14 dB to 10.5 dB). Further improvements require a redesign of the PIC. For example, changing the optical I/Os from GCs to edge couplers can reduce the cumulative I/O losses from 7.5 dB to 3 dB.

The critical components of the PIC have been measured after fabrication and show excellent agreement with modeling. A micrograph of the fabricated PIC is shown in Fig. 6.20, and the most relevant measurements are shown in Fig. 6.21 for the CROW OADMs and MZM phase shifter. After spectral alignment of the rings via the embedded PS, the 2nd- and 4th-order CROW OADMs have ILs of 3.5 dB and 2 dB and 3-dB bandwidths of 8 and 32 GHz, respectively (see Fig. 6.21a), on par with the expectations from modeling. The high-speed phase shifters of the IQ-MZMs achieve a $\frac{\pi}{2}$ phase shift at ~7 V when applying a reverse DC voltage, also in very good agreement with simulations, see Fig. 6.21b. When operated with 2 V_{pp} signals and biased at the minimum power point for QAM modulation, the IQ modulators introduce the expected IL and modulation penalty of 14 dB and have a BW in excess of 30 GHz, as shown in Fig. 6.21c.

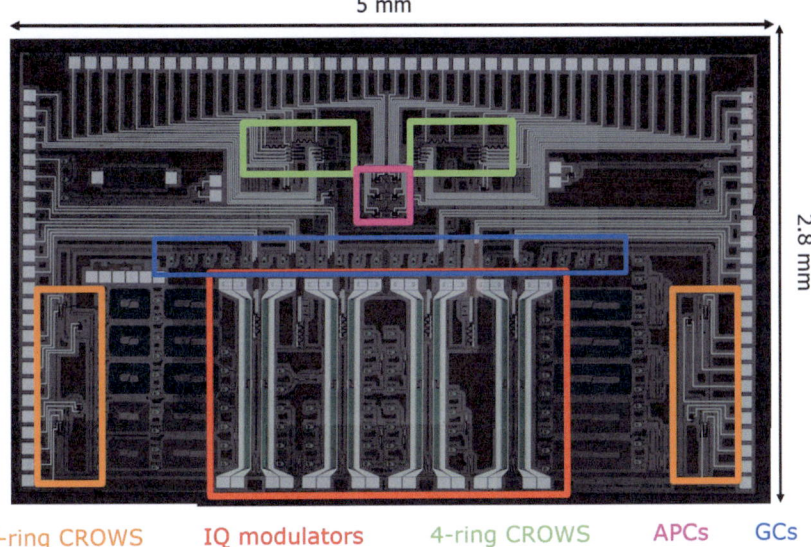

Fig. 6.20 Micrograph of the fabricated PIC

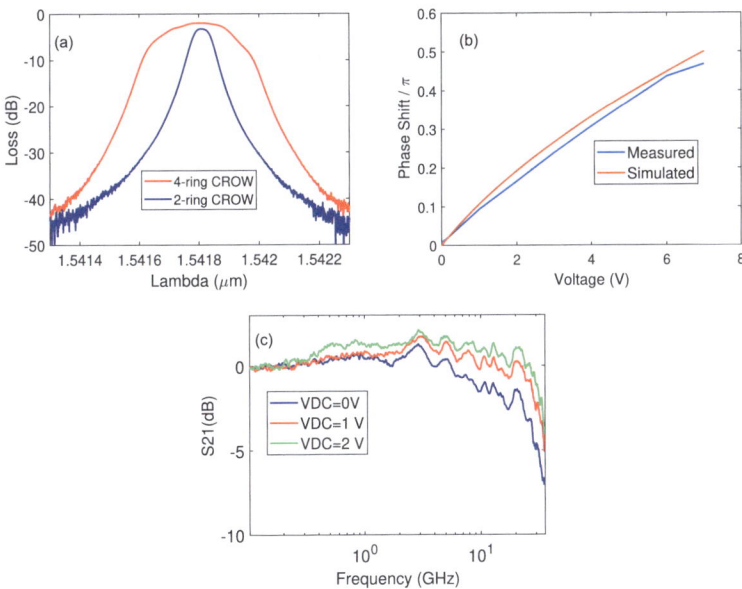

Fig. 6.21 a Transfer function to the CROW OADM's drop port after spectral alignment of the rings. b MZM phase shift when applying a DC voltage to a single arm. c Normalized S_{21} of the MZM

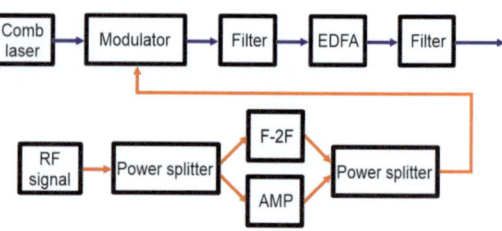

Fig. 6.22 Block diagram of the comb source generator

6.4.2 Comb Source Implementation and Measurements

Since only four coherent optical carriers are required, the comb source can be implemented using an E-O setup [59] following the block diagram from Fig. 6.22 and using off-the-shelf equipment. Light is emitted by a continuous wave tunable light source (Agilent 81960A) and injected in a 40 GHz BW Thorlabs LN05S lithium niobate modulator. The light source is modulated using two RF tones at 25 and 50 GHz to create five coherent carriers with a free spectral range (FSR) of 25 GHz. The center frequency and the FSR of the comb source can be easily adjusted by simply changing the optical/RF frequency of the laser and RF signal generator. To generate the 50 GHz RF tone, the 25 GHz signal is split into two arms using a power divider. The first arm consists of an RF amplifier (SHF S807 C) whose gain can be externally controlled to obtain a flat-top comb with equal power per line. The second arm consists of an active frequency doubler (Marki ADA-2052) that also amplifies the signal output, but without the ability to control its gain. After recombining both signals using a second power divider, the output is connected to the optical modulator. Once the coherent comb lines have been generated, external filters and an EDFA can be used to select and amplify the desired four optical lines.

The characteristics of the generated comb are shown in Fig. 6.23. It has four lines with an output power of 10 dBm each, which can be further amplified, an RF linewidth below 1 Hz that is beyond the ability of our spectrum analyzer to measure, and relative intensity noise (RIN) below -140 dBc/Hz at offset frequencies above 20 MHz. Since the optical linewidth of an E-O frequency comb is inherited from the used continuous wave laser source, an optical linewidth below 100 kHz is expected. The measurement shown in Fig. 6.23a is limited by the resolution bandwidth (RBW) of the optical spectrum analyzer (OSA).

6.4.3 RF and Control PCB Implementation

The RF printed circuit board (PCB) has been fabricated in a multi-layer stack that supports high-density routing of the RF and control signals (see Fig. 6.24). The RF transmission lines are implemented as grounded co-planar waveguides (GCPW) in the top two conductor layers, with the thin interlayer dielectric allowing to shrink the lateral dimensions of the GCPW and enabling a compact pad frame that matches that of the PIC. Control signals are

6 Stitched-Spectrum Nonlinear Frequency Division Multiplexed …

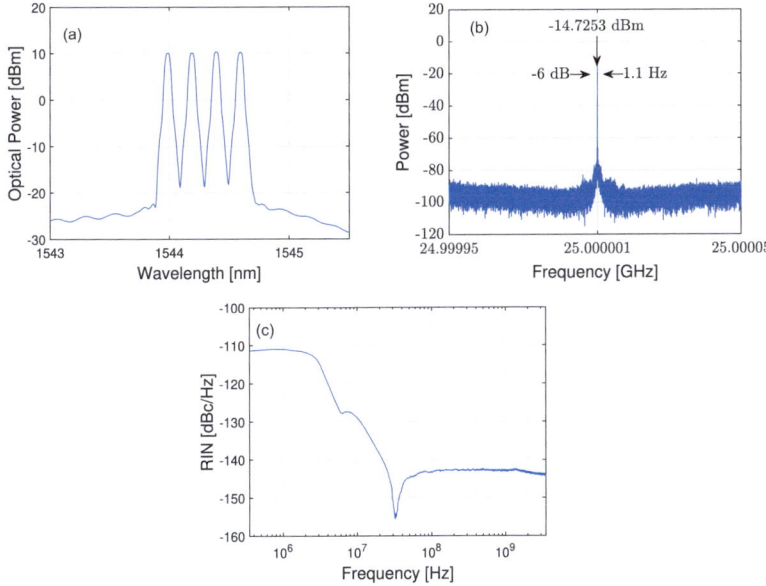

Fig. 6.23 **a** Optical spectrum, **b** RF spectrum and **c** RIN of the generated comb. The optical spectrum measurement in (**a**) is limited by the RBW of the OSA

Fig. 6.24 PCB multilayer stack concept

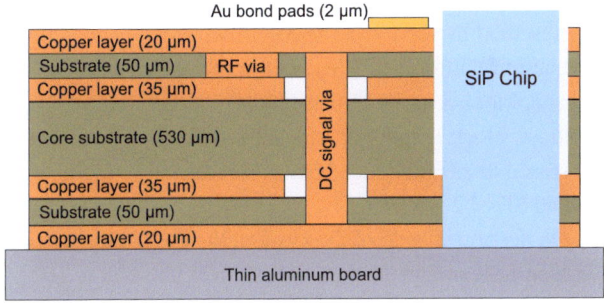

routed through the bottom two conductor layers. An opening provided in the PCB allows placing the PIC directly on the underlying heat sink, which is required due to the high heat-load of the PIC, that comprises a total of 39 thermally tuned phase shifters which require between 29 mW (PS embedded in the rings) and 44 mW (PS embedded in the MZMs) each to reach a 2π phase-shift and eight 50 Ω terminated high-speed phase shifters that consume 40 mW each when driven with a 2 V_{pp} signal. Since PSs are actuated to an average phase of π and only half of the PS, configured in push-pull configuration, are actuated in the IQ-modulators, the total power dissipated in the chip is ~1.4 W. In order to minimize the length of the wire-bonds, a thick substrate core is provided so that the surface of the PIC is flush with that of the PCB. The overall temperature of the PIC is controlled by a thermoelectric cooler (TEC) placed at the bottom of the aluminum board serving as a heat sink.

6.5 System Setup and Simulation Results

In this section, we model the performance of the proposed system taking into account transmitter and receiver non-idealities, i.e., the optical losses of the PIC, the excess amplified spontaneous emission (ASE) noise resulting from pre- and post-amplifying the light before and after the PIC, the RF and optical linewidths of the comb, the transfer functions of the OADMs, the cutoff frequency of the IQ-modulators, and the linewidth of the local oscillator laser in the receiver, in addition to the non-idealities already considered in Sect. 6.3.2.

A drawback of sharing equal data streams across even and odd channels is that dispersion pre-compensation cannot be applied. To determine how much this will handicap the experiments, we also run models with and without pre-compensation, wherein the former assumes a PIC with four independent IQ-modulators.

Figure 6.25 shows the schematic of the simulation setup used to assess the system capabilities. As previously, the system has an effective bandwidth (BW) of 100 GHz centered at 1550 nm, which is divided into four channels, each having a BW that can be adjusted between 25 and 32 GHz depending on the targeted spectral overlap. After the data is generated for each channel and mapped to 32-QAM, it follows the same digital processing flow as presented in Sect. 6.3. The electrical signals are again generated with an 88 GSa/s AWG and amplified to reach 2 V_{pp} before being injected into the PIC. When pre-compensation is used, block lengths of 6 ns result in a data rate of 333 Gbps and an SE of 3.3 b/s/Hz. In contrast, block lengths of 8 ns are required when no pre-compensation is used, due to the increased dispersion-based pulse broadening in the fiber, leading to a data rate of 250 Gbps and an SE of 2.5 b/s/Hz.

On the optical side of the system, the E-O frequency comb provides four optical carriers with an FSR of 25 GHz, an optical linewidth below 100 kHz, an RF linewidth below 1 Hz,

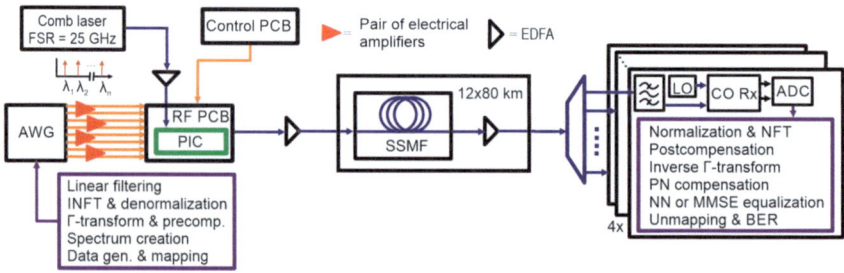

Fig. 6.25 Block diagram used for the simulation of the transmission system including the transmitter and receiver non-idealities

6 Stitched-Spectrum Nonlinear Frequency Division Multiplexed …

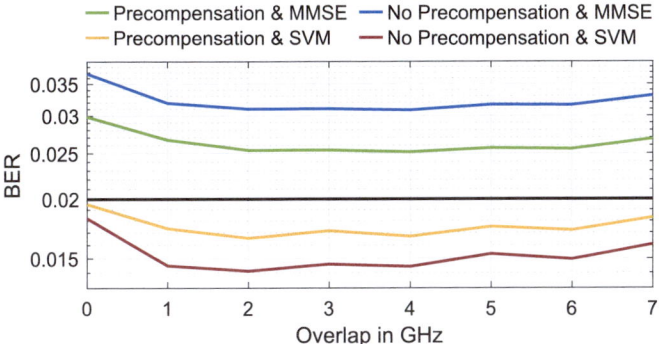

Fig. 6.26 BER versus spectral overlap for 32-QAM transmission, with the SD-FEC limit indicated in black. Comb lines are injected with 10 dBm into the PIC. Investigated scenarios include using or not using pre-compensation at the transmitter and using MMSE or SVM equalization at the receiver

and a power per line above 10 dBm. Since the comb is already pre-amplified, as shown in the diagram in Fig. 6.22, it would in principle be possible to amplify it further, so that the power is limited by the damage threshold of silicon-on-insulator (SOI) GCs instead, which is in the order of 18–20 dBm. The total comb power of 16 dBm is thus chosen to remain below it with an adequate safety margin. Given the overall PIC loss and modulation penalty of 33.5 dB (see Sect. 6.4.1), the multiplexed NFDM channels have an average power level of -23.5 dBm at the PIC output, which is re-amplified with a second EDFA by setting the output power to 0 dBm to reach the optimal launch power before being sent over 12 spans of 80 km SSMF as in the link configuration assumed in Sect. 6.3.

At the receiver, channels are again demultiplexed, filtered, and demodulated using a coherent receiver and an 80 GSa/s ADC. In the digital domain, the NFT, dispersion post-compensation, and inverse Γ-transform are computed. After that, phase noise (PN) compensation is applied using eight subcarriers in each channel as pilot tones. Before de-mapping and BER calculation, either an MMSE or a support vector machine (SVM) with a linear kernel is applied for equalization.

Figure 6.26 shows the simulated BER versus the spectral overlap, as obtained after applying MMSE and SVM equalization when the comb source is set to inject 10 dBm per line. There is a clear improvement once overlap of more than 1-2 GHz is applied, but the performance remains above the SD-FEC limit of 2×10^{-2} for MMSE equalization. Therefore, higher performing equalization techniques such as SVMs or neural networks (NN) are required. In general, the obtained results are slightly worse than the ones we previously presented in [41]. There are several reasons for this: The comb source we have finally implemented offers higher output power, but also a higher optical linewidth compared to the initially assumed one (100 kHz vs. 20 kHz); the PIC losses are 3.5 dB higher based on actual measurement data; and SVM with a linear kernel equalization is a much faster technique, but improves the BER less than NNs. It is interesting to note that the SVM-equalization

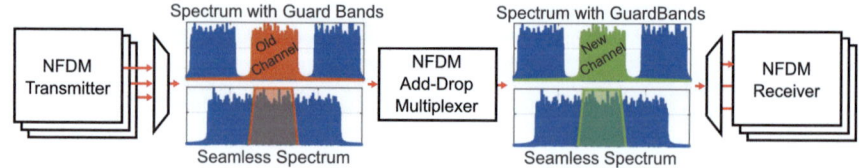

Fig. 6.27 Add-drop multiplexing for WDM NFDM transmission systems. The spectra in the top row represent the case of NFDM channels with guard bands, while the spectra in the bottom row represent the case of a seamlessly stitched spectrum

performs better in the absence of pre-compensation, similarly to what was observed for NN-equalization in [41], but in contrast to how MMSE-equalization behaves.

6.6 Add-Drop Multiplexing Concept

Routing is an essential capability that is required to send data to its intended destination in communication networks. In conventional WDM systems, this is typically done by using add-drop multiplexers, which are able to add or drop a channel at a certain point in the network. The application of this concept to NFDM is illustrated in Fig. 6.27. Multiple channels are created and multiplexed together in the NFDM transmitter. When using guard bands, as shown in the upper spectra, the function of the add-drop multiplexer is the same as for conventional systems: Since each channel is separate and independent from the others, it can easily be filtered out and a new channel can be added in the vacated space. However, in our case we generate a seamless spectrum, as shown in the lower spectra in Fig. 6.27, so that this method cannot be applied directly if only a portion of this spectrum is to be replaced. The channels are not independent from each other, so that information from the neighboring channels is needed to create and add a new channel. Due to this dependency, the new channel must also be placed in the same time slot as the dropped one, constraining the operation further.

In our concept, as for the transmitter described in the last sections, the new channel is added by operating a linear superposition with trapezoidal-shaped filter functions, to account for the nonlinear interaction. The phases of all optical signals have thus also to match to achieve a constructive superposition in the overlap regions.

6.6.1 Add-Drop Multiplexer Architecture

In the following, we present a concept for applying this add-drop multiplexing operation to a seamless NFDM spectrum, adapting the spectral overlap concept to this context in a manner that can also be supported by a PIC-based realization. The system architecture is an extension of an optical-electrical-optical (OEO) interferometer previously applied to a similar problem in the linear spectrum [48] and is shown in Fig. 6.28. It should be noted that for the NFDM receiver, it makes no difference if add-drop multiplexing has been applied or not, the signal is demultiplexed and demodulated in the same way as in the previous sections.

In the OEO interferometer, the incoming optical signal is split using a 3-dB splitter, after which one path is delayed using an optical delay line (ODL), e.g. an SSMF loop, while the other path is coupled into the PIC, where a CROW filter selects the channel to be replaced together with a portion of the surrounding spectrum as needed to be known to perform the nonlinear add-drop multiplexing operation, as explained in the following. A laser is provided as a local light source, whose output is also split to function as the local oscillator of a receiver used to analyze the incoming spectrum and as the carrier for a transmitter used to generate a signal that cancels and replaces the dropped channel. Once the spectrum selected by the CROW filter has been digitized, DSP is used to compute a signal that, when coherently combined with the original signal, will remove the target channel and add a new one. This signal is then converted back to the optical domain and amplified with an EDFA to reach the necessary launch power. A 2×2 directional coupler splitter (DCS) finally combines the new signal with the delayed original one. The delay introduced by the ODL that is applied to the initial optical signal before the DCS is matched to that of the DSP part, so that the resulting summation cancels out the channel to be replaced and coherently replaces it.

Since the coherent receiver and the IQ modulator share the same laser, its phase noise (PN) is first applied in a phase-conjugated manner during coherent detection, after which is it applied in a non-phase-conjugated manner during the optical signal generation, so that its PN can be canceled as a net result. For this to hold true, a second ODL applies a delay to the local laser, prior to feeding it to the IQ-modulator, that is also matched to the DSP part. Additional dynamic phase errors caused by the propagation through the two optical fiber loops can be corrected using a phase shifter (PS) cascaded with the output of the IQ modulator [45]. This phase correction is controlled by measuring the power of the combined signal at the complementary port of the DCS with a monitor photodiode (PD). Since the subtraction of the cancellation signal from the original one within the main optical signal path corresponds to a coherent summation at the monitor, the power detected by the PD has to be maximized by the control system.

Up to this point, the OEO interferometer closely resembles the architecture applied to linear systems, in which this fully accounts for phase noise issues. In our case, there is an additional difficulty associated to the phase noise of the laser used in the original transmitter that arises from applying this architecture to the NFDM framework, since now all the subcarriers are required to share a common phase noise as a consequence of the nonlinear interactions. This can be solved in the DSP part, as we shall see in the following.

In principle, the cancellation signal could be generated from the data transported by the dropped nonlinear channel. To recover the dropped channel taking nonlinear interaction into account, the analyzed linear spectrum, as provided at the drop port of the OADM, would then need to contain at least the channel with its two adjacent overlap regions, with some

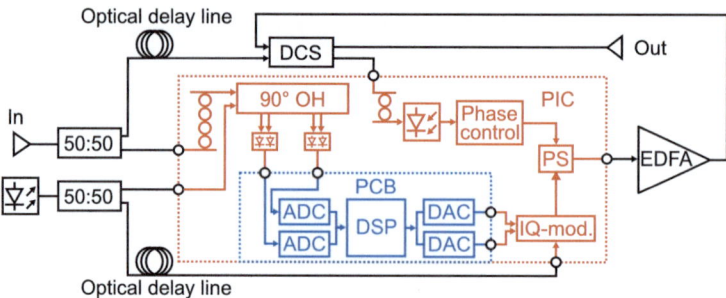

Fig. 6.28 Block diagram of the PIC-based add-drop module. Additional discrete components are shown in black, the PCB in blue, and the PIC in orange

additional subcarriers added from the adjacent channels to account for nonlinear interaction. The cancellation signal could then be generated from that data following the DSP flow of the transmitter. This approach may, however, be compromised by noise and attenuation occurring in the fiber link, the break the regularities of the NFT.

To circumvent this difficulty and facilitate data processing, the cancellation signal is directly generated from the recorded linear spectrum without using the NFT framework. For this simplified data processing flow to work without excess penalties, any region of the linear spectrum that contains spectral content of the dropped nonlinear channel needs to be fully extinguished. In order to achieve this, the width of the trapezoidal filters used in the OEO interferometer, that also defines the range in which the initial signal is extinguished, is widened compared to the one used in the transmitter, with the filters' flat-top region now also encompassing the initial overlap regions and with new overlap regions defined at their edges. This extended filter bandwidth configuration is illustrated in Fig. 6.30. The extended filter is applied to the recorded spectrum to generate the cancellation signal that is to be subtracted from the main signal path. Without this bandwidth extension, spectral content of the dropped channel would remain and would interfere with the new data.

Since the new overlap regions now extend into the adjacent channels, the synthetization of the added spectrum also needs to take subchannels of the adjacent channels into account in a wider range including the new overlap regions. These have thus first to be known and need to be extracted from the incoming signal. This further broadens the spectral range that has to be provided by the OADM multiplexers beyond the outer edges of the extended filters to adequately take nonlinear interaction into account. Additional steps are also necessary in the DSP part to recover these adjacent subchannels. Therefore, the whole receiver DSP has first to be applied as depicted in Fig. 6.29. The added spectrum is generated by applying the INFT to the added channel concatenated with the adjacent subcarriers of the adjacent channels and feeding the result to the trapezoidal filters with extended bandwidth.

Fig. 6.29 DSP chain for the PIC-based add-drop module

Fig. 6.30 Extended filter bandwidth configuration

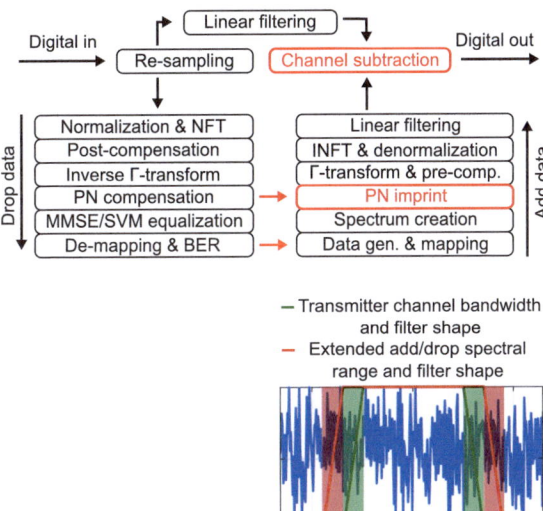

Another important point to consider is the PN of the original signal. Even if the PN of the local laser is adequately canceled out for the cancellation signal, the PN of the added signal has to be matched to that of the surrounding channels, so that coherent signal summation can occur in the overlap regions. This is not a problem in a conventional linear processing scheme in which added subcarriers are orthogonal to pre-existing ones. Here, however, subcarriers in the overlap regions are reconstructed via a coherent superposition of the pre-existing and added spectra. This is realized using the PN recorded from the PN compensation in the receiver part of the OEO interferometer, that comprises both the non-phase-conjugated PN of the original transmitter and the phase-conjugated PN of the local laser, and imprinting it on the new channel before modulation of the local laser. As a final step in the DSP part, after passing through the extended trapezoidal-shaped linear filters, the cancellation signal is subtracted from the added one prior to E-O transduction (Fig. 6.29).

6.6.2 Add-Drop Multiplexing with Continuous Spectrum Modulation

Figure 6.31 shows the simulation setup used to verify the signal quality after this nonlinear add-drop multiplexing operation. The system is similar to the one in Sect. 6.3, with the add-drop multiplexer added in the middle of the transmission link, and follows a similar level of modeling depth. As previously, the system has 4 channels with a spectral overlap of 4 GHz and a total bandwidth of 100 GHz. The channels are modulated with 32-QAM using b-modulation with the Γ-transform and dispersion pre-compensation applied, followed by an INFT and denormalization. As previously, after the trapezoidal-shaped filtering, an AWG with a sample rate of 88 GSa/s is used to generate the electrical signal applied by an IQ modulator to a line filtered out from the comb source. All channels are multiplexed together and the launch power is set to −0.5 dBm using an EDFA. After 5 spans of 80 km SSMF, the signal arrives at the add-drop multiplexer, in which channel 2 is replaced using the process described above, with the ADC and AWG also having sample rates of 80 GSa/s and 88 GSa/s, respectively. The signal is then sent through another 5 spans of 80 km SSMF and the channels are coherently received and digitized using an ADC with 80 GSa/s. The receiver DSP has the same characteristics as the one used for the add-drop multiplexer as shown in the left stack of Fig. 6.29. All optical sources have a linewidth of 100 kHz. Besides this and the noise figure of the EDFA, the OEO interferometer is modeled as an ideal system.

Figure 6.32 shows the EVM per subcarrier with and without PN imprint. It can be seen that the PN imprint is essential and that data cannot be transmitted in the overlap regions without it. This is expected, as explained above, because phase differences lead to random interference conditions in the overlap regions. The change in performance resulting from extending the add-drop filter bandwidths is depicted in Fig. 6.33. Here too, the stark change in performance at the edges of the channel is expected.

In an effort to increase the SE of the system, different numbers of subcarriers and different equalization schemes are evaluated with results shown in Fig. 6.34. On the left, 400 subcarriers are used as in previous configurations, resulting in a block length of 6 ns, a data rate of 333 Gbps and an SE of 3.33 b/s/Hz. MMSE equalization is sufficient for all subcarriers to stay below the SD-FEC limit of 2×10^{-2} and the BER can be further improved with a simple linear kernel SVM. In the right graph, 800 subcarriers are used instead, leading to increased block lengths of 10 ns ($T_c = 8$ ns), a data rate of 400 Gbps and an SE of 4 b/s/Hz. However, increasing the number of subcarriers decreases the signal quality [53], so that the MMSE equalization no longer manages to keep all the subcarriers below the SD-FEC limit. SVM equalization again improves the BER to a level below the SD-FEC limit and error-free transmission is possible.

These results show that the add-drop concept is sound and can be used to increase the flexibility of NFDM WDM transmission systems with seamless spectra. If, in addition to the continuous spectrum, the discrete spectrum is also to be used for data transmission, some additional important points have to be considered.

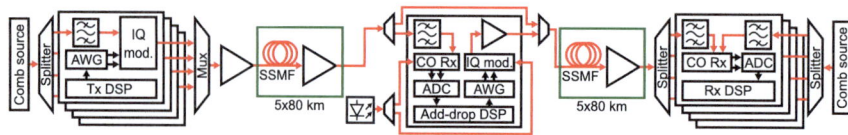

Fig. 6.31 Block diagram of the simulated transmission system including add-drop multiplexing

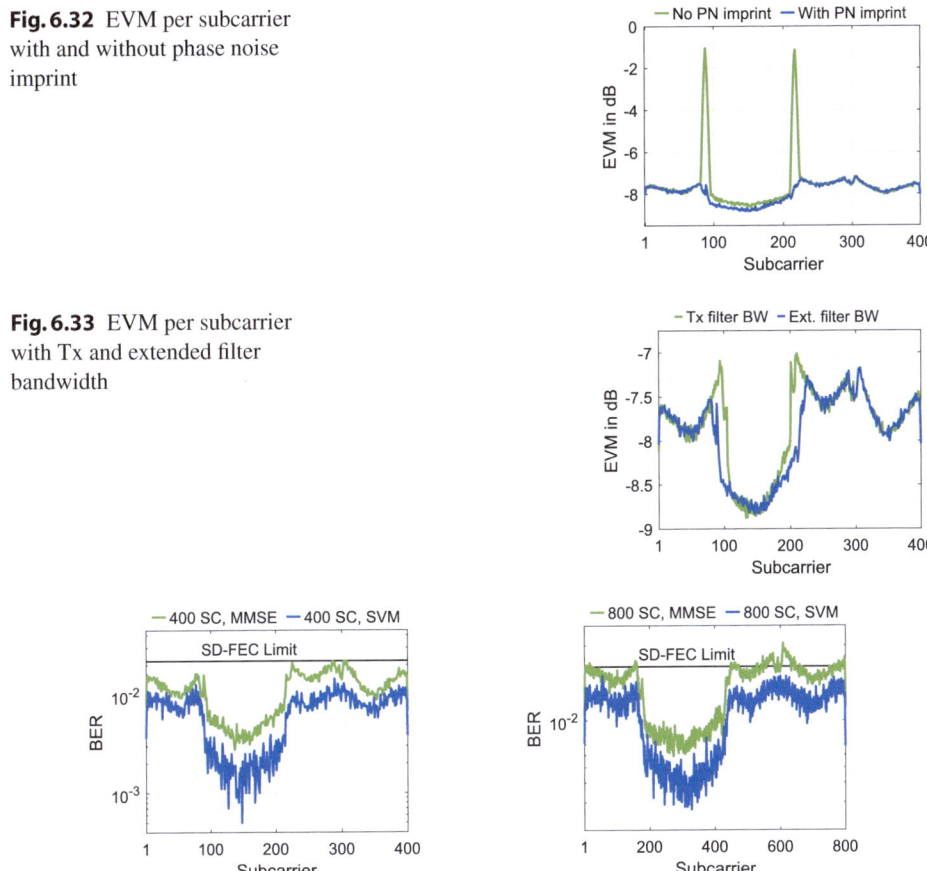

Fig. 6.32 EVM per subcarrier with and without phase noise imprint

Fig. 6.33 EVM per subcarrier with Tx and extended filter bandwidth

Fig. 6.34 BER per subcarrier with MMSE and SVM equalization using 400 or 800 subcarriers

6.6.3 Add-Drop Multiplexing with Full Spectrum Modulation

The simulation setup models the same hardware as before, with the continuous spectrum created to contain again a total of 400 subcarriers with 4 GHz overlap regions. The subcarriers are now modulated with 16-QAM, to which a discrete spectrum consisting of two eigenvalues per channel with $\Im(\lambda_j) = 1.75$, which proved to be optimum in Sect. 6.3.3, is added. However, in comparison to Sect. 6.3.3, more degrees of freedom are used to increase the SE: The amplitude (time position) $|b(\lambda_j)|$ is 16-ASK modulated, the phase $\angle b(\lambda_j)$ is 8-PSK modulated, and the soliton frequency $\Re(\lambda_j)$ is 32-FSK modulated, resulting in an overall modulation format of 4096-APFSK per eigenvalue. This leads to a total data rate of 282.67 Gbps and the SE is therefore increased from 2.72 b/s/Hz to 2.83 b/s/Hz. The

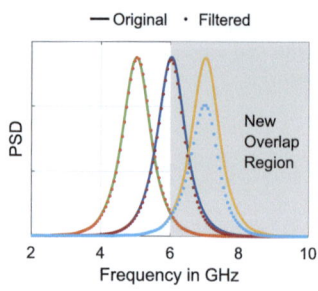

Fig. 6.35 Comparison of solitons partially overlaying the overlap region before and after trapezoidal-shaped filtering

equalization in the receiver is done using an SVM for the continuous spectrum and an MMSE for the discrete part.

In order for the frequency modulation of the eigenvalues to feature a high contrast, the frequency range should be as large as possible. However, it must be ensured that solitons are not positioned too close to the channel boundaries, as otherwise their spectrum will be excessively clipped by the trapezoidal-shaped filtering. This not only leads to the discrete spectrum, but also to the continuous spectrum to feature an increased BER, as this also compromises the orthogonality between the two. This is especially important to consider in the context of add-drop multiplexing, as the extended bandwidth of the filters further constrains the spectral range available to the soliton pulses of the adjacent channels. Even if the solitons are positioned adequately in the transmitter, this can again lead to a partial clipping in the add-drop multiplexer for channels adjacent to the one being replaced, as shown in Fig. 6.35, since the extended filter encroaches deeper into them. Moreover, the reconstruction of partially clipped soliton pulses in the add-drop multiplexer would require a very significant E-O bandwidth increase, as the channels adjacent to the one being replaced would then also need to be fully recovered. Consequently, the possibility of add-drop multiplexing must already be taken into account in the transmitter when designing the modulation format.

The resulting performance degradation is shown in Fig. 6.36, where the BERs of the different modulated spectral components of the NFT are plotted as a function of the highest eigenvalue-frequency modulation offsets. Here, only the BERs of the eigenvalues and the continuous spectrum subcarriers which are closest to the new overlap regions of the add-drop multiplexer are evaluated. It can be seen that especially the BER associated to the eigenvalue time position increases rapidly as soon as the soliton is influenced by the trapezoidal-shaped filtering. The BER of the other parts also increases, but more slowly, making them more robust against this.

A more detailed look at the performance of the continuous spectrum is given in Fig. 6.37. The BER remains below the assumed SD-FEC limit of 2×10^{-2} for all subcarriers for eigenvalue frequency offsets up to 6 GHz. At 7 GHz, the BER increases rapidly, making error-free transmission in the new overlap regions no longer possible. This makes the continuous spectrum robust against partial soliton truncation up to the point at which about half of the soliton pulse lies in the new overlap region. It can thus be seen that the add-drop multiplexer

Fig. 6.36 BER of the different modulated spectral components of the NFT for varying highest eigenvalue frequency modulation. CS: continous spectrum, OR: overlap region

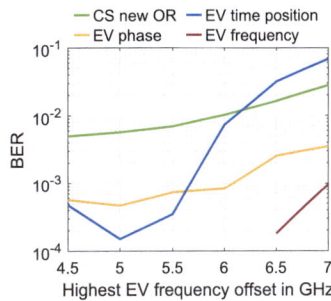

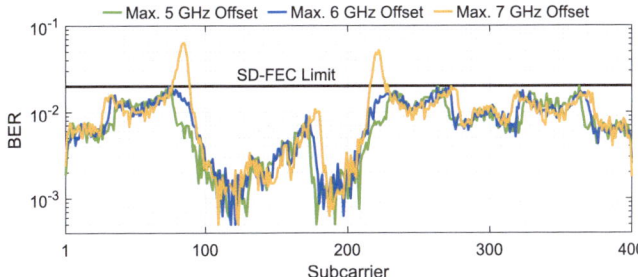

Fig. 6.37 Continuous spectrum BER for each subcarrier for different highest eigenvalue frequency modulation

concept is also applicable to full spectrum modulated systems, as long as the eigenvalues are positioned sufficiently far from the overlap regions during the entire transmission.

6.7 Conclusion and Outlook

We have presented a WDM NFDM transmission system which uses spectral overlap between adjacent channels to mitigate nonlinear inter-channel crosstalk. This highly scalable concept uses the whole available spectrum since no guard bands are required. We have numerically evaluated its performance with a 4-channel WDM system seamlessly filling a 100 GHz spectral range using both continuous and full spectrum modulation. Flexibility of the system is increased by an add-drop multiplexer concept which is compatible with the replacement of nonlinear channels in this seamless spectrum. The performance of the system was evaluated using different modulation formats and equalization schemes, showing that the system can achieve high spectral efficiencies of 4 b/s/Hz using the continuous spectrum with a single polarization over 800 km, applying a simple SVM equalization with a linear kernel. Further improvement may be possible by using neural network equalizers, dual polarization, and probabilistic shaping.

The WDM NFDM link has also been simulated taking the non-idealities of a fabricated transmitter PIC into account. As expected, the overall performance decreases a bit, but we achieved also here a BER remaining below the SD-FEC limit with SVM equalization using the 400 subcarrier configuration (SE of 3.33 b/s/Hz).

The concepts described here are not only applicable to the framework of the nonlinear Fourier transform, but to any optical signal processing scheme in which wide seamless spectra are to be launched with nonlinear pre-compensation. As such, they may be instrumental in increasing the data throughput beyond the current limitations of the nonlinear Shannon limit in future communication systems in a more general nonlinear signal processing and synthesis context.

Acknowledgements The authors would like to acknowledge funding from the German Research Foundation (Deutsche Forschungsgemeinschaft–DFG) for project "Nonlinear Fourier Transform based Optical Transmission using Electronic-Photonic Signal Processing" (403153975) in the framework for the Priority Programme 2111 "Electronic-Photonic Integrated Systems for Ultrafast Signal Processing".

References

1. Essiambre R-J, Kramer G, Winzer PJ et al (2010) Capacity limits of optical fiber networks. J Lightwave Technol 28(4):662–701. https://doi.org/10.1109/JLT.2009.2039464
2. Puttnam BJ, Rademacher G, Luís RS (2021) Space-division multiplexing for optical fiber communications. Optica 8(9):1186–1203. https://doi.org/10.1364/OPTICA.427631
3. Turitsyn SK, Prilepsky JE, Le ST et al (2017) Nonlinear Fourier transform for optical data processing and transmission: advances and perspectives. Optica 4(3):307–322. https://doi.org/10.1364/OPTICA.4.000307
4. Gui T, Lu C, Lau APT et al (2017a) High-order modulation on a single discrete eigenvalue for optical communications based on nonlinear Fourier transform. Opt Express 25(17):20286–20297. https://doi.org/10.1364/OE.25.020286
5. Gui T, Chan TH, Lu C et al (2017b) Alternative decoding methods for optical communications based on nonlinear Fourier transform. J Lightwave Technol 35(9):1542–1550. https://doi.org/10.1109/JLT.2017.2654493
6. Nakazawa M, Kubota H, Suzuki K et al (2000) Recent progress in soliton transmission technology. Chaos 10(3):486–514. https://doi.org/10.1063/1.1311394
7. Gaiarin S, Perego AM, da Silva EP et al (2018) Dual-polarization nonlinear Fourier transform-based optical communication system. Optica 5(3):263–270. https://doi.org/10.1364/OPTICA.5.000263
8. Moscoso-Mártir A, Koch J, Schulz O et al (2022) Silicon photonic integrated circuits for soliton based long haul optical communication. J Lightwave Technol 40(10):3210–3222. https://doi.org/10.1109/JLT.2022.3170250
9. Yousefi MI, Kschischang FR (2014) Information transmission using the nonlinear Fourier transform, part I-III. IEEE Trans Inf Theory 60(7):4312–4369. https://doi.org/10.1109/TIT.2014.2321143

10. Le ST, Aref V, Buelow H (2018b) High speed precompensated nonlinear frequency-division multiplexed transmissions. J Lightwave Technol 36(6):1296–1303. https://doi.org/10.1109/JLT.2017.2787185
11. Yangzhang X, Aref V, Le ST et al (2018) 400 Gbps dual-polarisation non-linear frequency-division multiplexed transmission with b-modulation. In: Proceedings of the 44th European conference on optical communication (ECOC), Rome, Italy, 23–27 Sept 2018. https://doi.org/10.1109/ECOC.2018.8535128
12. Le ST, Aref V, Buelow H (2017) Nonlinear signal multiplexing for communication beyond the Kerr nonlinearity limit. Nat Photonics 11:570–576. https://doi.org/10.1038/nphoton.2017.118
13. Goossens J-W, Yousefi MI, Jaouën Y et al (2017) Polarization-division multiplexing based on the nonlinear Fourier transform. Opt Express 25(22):26437–26452. https://doi.org/10.1364/OE.25.026437
14. Le ST, Schuh K, Buchali F et al (2018a) 100 Gbps b-modulated nonlinear frequency division multiplexed transmission. In: Proceedings of the optical fiber communications conference and exposition (OFC), San Diego, USA, 11–15 Mar 2018. https://doi.org/10.1364/OFC.2018.W1G.6
15. Yangzhang X, Le ST, Aref V et al (2019a) Experimental demonstration of dual-polarization NFDM transmission with b-modulation. IEEE Photonics Technol Lett 31(11):885–888. https://doi.org/10.1109/LPT.2019.2911600
16. Zhang Q, Kschischang FR (2021) Correlation-aided nonlinear spectrum detection. J Lightwave Technol 39(15):4923–4931. https://doi.org/10.1109/JLT.2021.3078700
17. Aref V, Le ST, Buelow H (2016) Demonstration of fully nonlinear spectrum modulated system in the highly nonlinear optical transmission regime. In: Proceedings of the 42nd European conference on optical communication (ECOC), Dusseldorf, Germany, 18–22 Sept 2016. https://doi.org/10.48550/arXiv.1611.08420
18. Aref V, Le ST, Buelow H (2018) Modulation over nonlinear Fourier spectrum: continuous and discrete spectrum. J Lightwave Technol 36(6):1289–1295. https://doi.org/10.1109/JLT.2018.2794475
19. Zhang R, Xi L, Zhang X et al (2022b) First demonstration of 113 Gb/s full-spectrum modulated nonlinear frequency division multiplexing transmission system. In: Proceedings of the conference on lasers and electro-optics (CLEO), San Jose, USA, 15–20 May 2022. https://doi.org/10.1364/CLEO_SI.2022.SM2J.5
20. Wei J, Xi L, Zhang X et al (2022) Improvement for a full-spectrum modulated nonlinear frequency division multiplexing transmission system. Opt Express 30(17):31195–31208. https://doi.org/10.1364/OE.465574
21. Pankratova M, Vasylchenkova A, Derevyanko SA et al (2020) Signal-noise in optical-fiber communication systems employing nonlinear frequency-division multiplexing. Phys Rev Appl 13(5):054021. https://doi.org/10.1103/PhysRevApplied.13.054021
22. Zhang WQ, Chan TH, Vahid SA (2022a) Serial and parallel convolutional neural network schemes for NFDM signals. Sci Rep 12:7962. https://doi.org/10.1038/s41598-022-12141-4
23. Chen X, Ming H, Li C et al (2021a) Two-stage artificial neural network-based burst-subcarrier joint equalization in nonlinear frequency division multiplexing systems. Opt Lett 46(7):1700–1703. https://doi.org/10.1364/OL.422195
24. Kotlyar O, Kamalian-Kopae M, Pankratova M et al (2021) Convolutional long short-term memory neural network equalizer for nonlinear Fourier transform-based optical transmission systems. Opt Express 29(7):11254–11267. https://doi.org/10.1364/OE.419314
25. Kotlyar O, Pankratova M, Kamalian-Kopae M et al (2020) Combining nonlinear Fourier transform and neural network-based processing in optical communications. Opt Lett 45(13):3462–3465. https://doi.org/10.1364/OL.394115

26. Zimu L, Wang Y, Han L et al (2024) CNN based equalizer in NFDM system with b-modulation. In: Proceedings of the 22nd international conference on optical communications and networks (ICOCN), Harbin, China, 26–29 July 2024. https://doi.org/10.1109/ICOCN63276.2024.10648413
27. Wei J, Xi L, Zhang X et al (2021) Probabilistic shaping and neural network-based optimization for a nonlinear frequency division multiplexing system. Opt Lett 46(15):3697–3700. https://doi.org/10.1364/OL.430859
28. Li C, Chen X, Chen Z et al (2021) Capacity increase in dual-polarization nonlinear frequency division multiplexing systems with probabilistic shaping. In: Proceedings of the 26th optoelectronics and communications conference (OECC), Hong Kong, Hong Kong, 3–7 July 2021. https://doi.org/10.1364/OECC.2021.W1B.2
29. Balogun M, Derevyanko S (2022a) Hermite-gaussian nonlinear spectral carriers for optical communication systems employing the nonlinear Fourier transform. IEEE Commun Lett 26(1):109–112. https://doi.org/10.1109/LCOMM.2021.3121541
30. Balogun M, Derevyanko S (2022b) Enhancing the spectral efficiency of nonlinear frequency division multiplexing systems via Hermite-gaussian subcarriers. J Lightwave Technol 40(18):6071–6077. https://doi.org/10.1109/JLT.2022.3188577
31. Balogun M, Barry L, Derevyanko S (2023) Enhanced achievable information rate for NFDM systems using FBMC wave-carriers. In: Proceedings of the 49th European conference on optical communications (ECOC), Glasgow, UK, 1–5 Oct 2023. https://doi.org/10.1049/icp.2023.2254
32. Chen X, Zhang F, Zhang X et al (2022a) 20.48 Tb/s over 1200km WDM transmission with nonlinear frequency division multiplexing. In: Proceedings of the Asia communications and photonics conference (ACP), Shenzhen, China, 5–8 Nov 2022. https://doi.org/10.1109/ACP55869.2022.10088853
33. Chen X, Fang X, Cai X et al (2023) Terabit ultra-high-speed NFDM transmission over G.654E fiber with 103Gbaud probabilistic shaped 64QAM signals. In: Proceedings of the conference on lasers and electro-optics (CLEO), San Jose, USA, 7–12 May 2023. https://doi.org/10.1364/CLEO_SI.2023.SM3I.1
34. Chimmalgi S, Prins PJ, Wahls S (2019) Fast nonlinear Fourier transform algorithms using higher order exponential integrators. IEEE Access 7:145161–145176. https://doi.org/10.1109/ACCESS.2019.2945480
35. Medvedev SB, Kachulin DI, Chekhovskoy IS et al (2024) New fast exponential splitting schemes for nonlinear Fourier transform. In: Proceedings of the 21st international conference laser optics (ICLO), Saint Petersburg, Russian Federation, 1–5 July 2024. https://doi.org/10.1109/ICLO59702.2024.10624105
36. Chen X, Fang X, Yang F et al (2021b) 6.4Tb/s (16x400Gb/s) nonlinear frequency division multiplexing WDM transmission over 640km SSMF. In: Proceedings of the 26th optoelectronics and communications conference (OECC), Hong Kong, Hong Kong, 3–7 July 2021. https://doi.org/10.1364/OECC.2021.T5A.4
37. Chen X, Fang X, Yang F et al (2022b) 10.83 Tb/s over 800 km nonlinear frequency division multiplexing WDM transmission. J Lightwave Technol 40(16):5385–5394. https://doi.org/10.1109/JLT.2022.3177413
38. Fang D, Zazzi A, Müller J et al (2021) Optical arbitrary waveform measurement (OAWM) using silicon photonic slicing filters. J Lightwave Technol 40(6):1705–1717. https://doi.org/10.1109/JLT.2021.3130764
39. Henauer T, Sherifaj A, Füllner C et al (2022) 200 GBd 16QAM signals synthesized by an actively phase-stabilized optical arbitrary waveform generator (OAWG). In: Proceedings of the optical fiber communications conference and exhibition (OFC), San Diego, USA, 6–10 Mar 2022. https://doi.org/10.1364/OFC.2022.M2I.2

40. Chen X, Fang X, Yang F et al (2021c) All-optical synthesis of 100Gbaud PDM 16-QAM nonlinear frequency division multiplexing signal. In: Proceedings of the Asia communications and photonics conference (ACP), Shanghai, China, 24–27 Oct 2021. https://doi.org/10.1364/ACPC.2021.T4D.5
41. Moscoso-Mártir A, Schulz O, Misra A et al (2023) Spectrally stitched WDM nonlinear frequency division multiplexed transmission system. Optics Commun 546:129809. https://doi.org/10.1016/j.optcom.2023.129809
42. Schulz O, Moscoso-Mártir A, Witzens J et al (2023a) Full spectrum WDM nonlinear frequency division multiplexed transmission system using spectral overlap. In: Proceedings of the advanced photonics congress (SPPCom), Busan, Republic of Korea, 9–13 July 2023. https://doi.org/10.1364/SPPCOM.2023.SpW3E.3
43. Schulz O, Moscoso-Mártir A, Witzens J et al (2023b) Highly scalable WDM nonlinear frequency division multiplexed transmission system using spectral overlap. In: Proceedings of the optical fiber communications conference and exhibition (OFC), San Diego, USA, 5–9 Mar 2023. https://doi.org/10.1364/OFC.2023.Th1F.4
44. Mahmud MS, Kemal JN, Adib M et al (2020) Optic-electronic-optic interferometer: a first experimental demonstration. In: Proceedings of the conference on lasers and electro-optics (CLEO), Washington, USA, 10–16 May 2020. https://doi.org/10.1364/CLEO_SI.2020.SF1L.1
45. Mahmud MS, Matalla P, Adib MMH et al (2023) Coherent add/drop multiplexing using an optic-electronic-optic interferometer. In: Proceedings of the conference on lasers and electro-optics (CLEO), San Jose, USA, 7–12 May 2023. https://doi.org/10.1364/CLEO_SI.2023.SM2I.6
46. Schulz O, Moscoso-Mártir A, Witzens J et al (2024b) Add-drop multiplexing for spectrally overlapped nonlinear frequency division multiplexed transmission systems. In: Proceedings of the optical fiber communications conference and exhibition (OFC), San Diego, USA, 24–28 Mar 2024. https://doi.org/10.1364/OFC.2024.W2A.26
47. Schulz O, Moscoso-Mártir A, Witzens J et al (2024a) Add-drop multiplexing for full spectrum WDM NFDM transmission systems using spectral overlap. In: Proceedings of the advanced photonics congress (SPPCom), Québec City, Canada, 28 July–1 Aug 2024. https://doi.org/10.1364/SPPCOM.2024.SpTh1H.2
48. Winzer PJ (2013) An opto-electronic interferometer and its use in subcarrier add/drop multiplexing. J Lightwave Technol 31(11):1775–1782. https://doi.org/10.1109/JLT.2013.2257687
49. Zakharov VE, Shabat AB (1972) Exact theory of two-dimensional self-focusing and one-dimensional self-modulation of waves in nonlinear media. Sov J Exp Theor Phys 34(1):62–69
50. Yousefi M, Yangzhang X (2020) Linear and nonlinear frequency-division multiplexing. IEEE Trans Inf Theory 66(1):478–495. https://doi.org/10.1109/TIT.2019.2941479
51. Prilepsky JE, Derevyanko SA, Blow KJ et al (2014) Nonlinear inverse synthesis and eigenvalue division multiplexing in optical fiber channels. Phys Rev Lett 113:013901. https://doi.org/10.1103/PhysRevLett.113.013901
52. Wahls S (2017) Generation of time-limited signals in the nonlinear Fourier domain via b-modulation. In: Proceedings of the 43rd European conference on optical communication (ECOC), Gothenburg, Sweden, 17–21 Sept 2017. https://doi.org/10.1109/ECOC.2017.8346231
53. Yangzhang X, Aref V, Le ST et al (2019b) Dual-polarization non-linear frequency-division multiplexed transmission with b-modulation. J Lightwave Technol 37(6):1570–1578. https://doi.org/10.1109/JLT.2019.2902961
54. Derevyanko S, Balogun M, Aluf O et al (2021) Channel model and the achievable information rates of the optical nonlinear frequency division-multiplexed systems employing continuous b-modulation. Opt Express 29(5):6384–6406. https://doi.org/10.1364/OE.414885
55. Tavakkolnia I, Safari M (2016) Dispersion pre-compensation for NFT-based optical fiber communication systems. In: Proceedings of the conference on lasers and electro-optics (CLEO), San Jose, USA, 5–10 June 2016. https://doi.org/10.1364/CLEO_SI.2016.SM4F.4

56. Vaibhav V (2018) Fast inverse nonlinear Fourier transform. Phys Rev E 98:013304. https://doi.org/10.1103/PhysRevE.98.013304
57. Fischer JK, Winter M, Petermann K (2009) Scaling of nonlinear threshold with fiber type and channel spacing in WDM transmission systems. In: Proceedings of the optical fiber communications conference (OFC), San Diego, USA, 22–26 Mar 2009. https://doi.org/10.1364/NFOEC.2009.JThA41
58. Moscoso-Mártir A, Tabatabaei-Mashayekh A, Müller J et al (2018) 8-channel WDM silicon photonics transceiver with SOA and semiconductor mode-locked laser. Opt Express 26(19):25446–25459. https://doi.org/10.1364/OE.26.025446
59. Zhuang R, Ni K, Wu G et al (2023) Electro-optic frequency combs: theory, characteristics and applications. Laser Photonics Rev 17(6):2200353. https://doi.org/10.1002/lpor.202200353
60. Deniel L, Weckenmann E, Galacho DP et al (2021) Silicon photonics phase and intensity modulators for flat frequency comb generation. Photonics Res 9(10):2068–2076. https://doi.org/10.1364/PRJ.431282
61. Pfeifle J, Brasch V, Lauermann M et al (2014) Coherent terabit communications with microresonator Kerr frequency combs. Nat Photonics 8:375–380. https://doi.org/10.1038/nphoton.2014.57
62. Siew SY, Li B, Gao F et al (2021) Review of silicon photonics technology and platform development. J Lightwave Technol 39(13):4374–4389. https://doi.org/10.1109/JLT.2021.3066203
63. Merget F, Azadeh S, Müller J et al (2013) Silicon photonics plasma-modulators with advanced transmission line design. Opt Express 21(17):19593–19607. https://doi.org/10.1364/OE.21.019593
64. Azadeh SS, Müller J, Merget F et al (2014) Advances in silicon photonics segmented electrode Mach-Zehnder modulators and peaking enhanced resonant devices. In: Proceedings of photonics North 2014, Montréal, Canada, 28–30 May 2014. https://doi.org/10.1117/12.2075836
65. Witzens J (2018) High-speed silicon photonics modulators. Proc IEEE 106(12):2158–2182. https://doi.org/10.1109/JPROC.2018.2877636
66. Müller J, Zazzi A, Rajeswari GV et al (2021) Optimized hourglass-shaped resonators for efficient thermal tuning of CROW filters with reduced crosstalk. In: Proceedings of the 17th international IEEE conference on group IV photonics (GFP), Malaga, Spain, 7–10 Dec 2021. https://doi.org/10.1109/GFP51802.2021.9673904
67. Yang Y, Ma Y, Guan H et al (2015) Phase coherence length in silicon photonic platform. Opt Express 23(13):16890–16902. https://doi.org/10.1364/OE.23.016890

Open Access This chapter is licensed under the terms of the Creative Commons Attribution 4.0 International License (http://creativecommons.org/licenses/by/4.0/), which permits use, sharing, adaptation, distribution and reproduction in any medium or format, as long as you give appropriate credit to the original author(s) and the source, provide a link to the Creative Commons license and indicate if changes were made.

The images or other third party material in this chapter are included in the chapter's Creative Commons license, unless indicated otherwise in a credit line to the material. If material is not included in the chapter's Creative Commons license and your intended use is not permitted by statutory regulation or exceeds the permitted use, you will need to obtain permission directly from the copyright holder.

If you have any concerns about our products,
you can contact us on
ProductSafety@springernature.com

In case Publisher is established outside the EU,
the EU authorized representative is:
**Springer Nature Customer Service Center GmbH
Europaplatz 3, 69115 Heidelberg, Germany**

Printed by Libri Plureos GmbH
in Hamburg, Germany